EMILIO AMBASZ
EMERGING NATURE

WITHDRAWN BY THE
UNIVERSITY OF MICHIGAN

EMILIO AMBASZ

Precursor of Architecture and Design

EMERGING NATURE

Essays by
Barry Bergdoll
Peter Buchanan
Kenneth Frampton
Peter Hall
Fulvio Irace
Dean MacCannell
Lauren Sedofsky

Interviews with
Hans Ulrich Obrist
Michael Sorkin
James Wines

Lars Müller Publishers

7	**THE RATIONALITY OF THE IMPROBABLE**
	Barry Bergdoll
22	La Casa de Retiro Espiritual, Seville, Spain, 1975
39	**THE ARCHITECTURE OF EMILIO AMBASZ**
	Peter Buchanan
52	Lucile Halsell Conservatory, San Antonio Botanical Garden, San Antonio, Texas, USA, 1982
65	**PERIPHERAL VISION**
	Lauren Sedofsky
88	Fukuoka Prefectural International Hall, Fukuoka, Japan, 1990
103	**IDEOLOGICAL CASTLES**
	Dean MacCannell
122	Mycal Cultural and Athletic Center, Shin-Sanda, Japan, 1990
137	**ARGENTINIAN AESOP**
154	**VISUAL APHORISMS**
	Fulvio Irace
156	**COOPERATIVE OF MEXICAN-AMERICAN GRAPE GROWERS**
160	**EMILIO'S FOLLY: MAN IS AN ISLAND**
	Emilio Ambasz

- 162 GREEN FACADE VERTICAL GARDEN
- 166 THE GREEN MOUNTAIN
- 170 THE EARTH AS A GARDEN
- 174 BUILDING IN THE GARDEN
- 177 URBAN GARDENS
- 179 DOMESTIC GARDENS
- 182 GARDENS OF MEMORY
- 184 GARDENS OF HEALTH
  Fulvio Irace

- 186 Ospedale dell'Angelo, Venice-Mestre, Italy, 2008
- 198 Banca degli Occhi, Venice-Mestre, Italy, 2009

- 207 THE IMMORTAL
  Peter Hall

- 220 Industrial and Graphic Design

- 237 INTERVIEW   Hans Ulrich Obrist with Emilio Ambasz

- 254 *Italy: The New Domestic Landscape,* Exhibition, 1972
- 262 *The Taxi Project: Realistic Solutions for Today,* Exhibition, 1976

- 267 THE ARCHITECT AS IDEAS MAN: A MEMOIR / CRITIQUE
  Kenneth Frampton

- 281 REPLIES   to Michael Sorkin's Questions
- 287 REPLIES   to James Wines's Questions

- 290 The Museum of Architectonic and Design Arts, Madrid, Spain, 2011–2012

- 301 Emilio Ambasz, Curriculum Vitae
- 304 Authors' Biographies
- 306 Index

A poetic realism lies at the heart of Ambasz's work and life, ever provocative, ever hard to pin down.

Barry Bergdoll

THE RATIONALITY
OF THE IMPROBABLE

Some notes on Emilio Ambasz's work in dialogue
with a spiritual retreat

Barry Bergdoll

How does one feel at home with, or even begin—as an historian—to situate the work of an architect whose most autobiographical work is a house whose facades house no rooms and whose rooms are almost without facades? What is to be made of a house with an earth-covered series of bedrooms and reception rooms, all removed from the commanding views of an estate of 600 hectares set in the verdant hills of the Sierra Morena outside Seville. Here the astounding natural setting can be appreciated only from a screened balcony reached by the most precipitous climb via one of the twin, hand-railing-free, cantilevered staircases anchored in the monumental walls that once inside reveal themselves to be but a staging of the *idea* of a house? How does one fathom such a noncanonic domestic gestalt for an owner/designer who claims never to have slept in his dream house? And how does one explain a house which few have ever seen, but which has been the subject of exhibitions, a lavish monograph, a dedicated website, and a host of prescient critiques? And to what historical context should we attach a house that helped bring fame to an untried architect when the project won a prestigious *Progressive Architecture* project award in 1975, even if a spate of interpretation would arise only a quarter century later, when the house was finally constructed hinting at its continued key relevance to his larger architectural philosophy? Yet just what is the meaning of an architect's house on a remote site in Andalusia when its owner/nonoccupant lives primarily between Bologna, Venice, and New York, only visiting his Casa de Retiro Espiritual from time to time—and rarely for more than a few hours—even as he continues to add other structures to its grounds? More questions perhaps than answers, these are nonetheless seeming contradictions that suggest possible points of entry into the world of Emilio Ambasz, the door of whose Spanish house first appears to open into a void. His is a world constructed perhaps less by buildings than by fables—as he himself proclaims—even as the published portfolio of his work (built and unbuilt) burgeons further each year with projects that oscillate between technological inventions and fictional devices. This is the dual universe created by both designer and designs, a world of allusion and elusiveness in which logical design choices join up over and over again with unprecedented imagery, unexpected images which nonetheless seem to conjure up a world of archetypes. In the quest for answers insights often arise as resolution remains elusive. A poetic realism lies at the heart of Ambasz's work and life, ever provocative, ever hard to pin down.

Like another icon of the twentieth century, Mies van der Rohe's 1929 Barcelona Pavilion, the Casa de Retiro Espiritual is at once highly site-specific and yet a mental construct that leaves powerful afterimages in the mind. Its contradictory forms are so powerful that images of the house alone fuel the imagination, nourish discussion, and transport us quickly beyond the elusive and allusive nature of the specific project to reflections about dwelling, about man and nature, and about architecture in general. Like the Barcelona Pavilion, another "house" that was never occupied and a building inhabited as much through images as experience, the Casa de Retiro Espiritual stands both as a manifesto and an accomplished statement of a body of work, a work at once exceptional in its enigmatic qualities and yet a kind of Rosetta stone for interpreting Ambasz's extensive practice. And like the huge body of interpretations of the Barcelona Pavilion, each successive interpretation of Ambasz's house layers further the richness of the work rather than displacing earlier interpretations. The Casa de Retiro Espiritual invites us to a restful mental sojourn without arduous travel, to the quiet contemplation of its contrasts and often irresolvable paradoxes; it serves as the perfect vestibule to an exploration of the real world sites of Ambasz's built work from Texas to Japan, and of his projected work from the California desert to Monte Carlo.

Like a fable the house has both a temporal and an atemporal existence. And since a quarter century separates design from construction, the project belongs also to different moments in both Ambasz's career and in the culture of architecture. It was designed in 1975, even before there was a site for the fantasy of a future habitat, projected for an unspecified couple. Then the young Argentina-born architect achieved precocious renown in New York with a series of brilliant exhibitions and conferences at the Museum of Modern Art in which he set out to call into question everything from the "domestic landscape" to the philosophical underpinnings of such major institutions of societal reproduction as the museum and the university. Likewise, his own house sets up philosophical conundrums for anyone who encounters it, in reality or via representations, just as the project remains an unfulfilled destination for the designer more than a decade after he has finally rendered a dreamed project literally concrete. The House of Spiritual Retreat seems more to be a memory palace, a spatial figure in the mind that helps order thoughts, recollections, memories, and forms over time — as it was imagined by the ancients and commemorated in a seminal and influential book by Frances Yates, *The Art of Memory* (1966). The house—if indeed it even is a house—seems almost to be a mechanism for rearranging the canonical in such a way as to allow memories to become productive rather than merely retrospective, a device that recalls the work of literary and artistic surrealisms, where archetypes seem to recover not so much lost traditions as lost possibilities.

Although the house seems to call out for some historical situating—if only at the simplest level to relate its forms to the vernacular history of the southern Spanish house whose patios, Islam-derived mashrabiya, and whitewashed walls all seem both honored and undermined in the design—it is interesting to recall the context of 1970s New York as a crucible rather than dwelling on the "timeless" vernacular of Andalucía. In 1975, with the "House of Spiritual Retreat" a thirty-two-year-old Argentinian curator at New York's Museum of Modern Art—having completed an architectural degree at Princeton in record time and captured the discursive limelight at twenty-nine with the landmark exhibition extravaganza *Italy: The New Domestic Landscape* (1972)—created a design that could dialogue with architectural masters and debates in multiple arenas. The design was a first episode in a lifelong conversation with Ambasz's spiritual mentor and first employer, the Argentinian engineer-architect Amancio Williams. This dialogue, begun in maquette, was continued in 1990 with an imagined epistolary relationship "Algunas notas sobre una correspondencia mental que mantuve a través de los últimos veinticinco años con Delfina Gálvez sobre la obra de Amancio," a series of letters depicting twenty years but composed at the then-present. At the same time the house was clearly a retrospective response to the theoretical architectural statements emanating from the Italian neo-avant-garde exhibited in 1972 in *Italy: The New Domestic Landscape,* such as Superstudio and Archizoom, in whose imagery the possibility of inhabiting Earth had been called into question. Whereas these Italian radicals, deeply tinged with nihilism, projected a future light inhabitation of the surface of an unforgiving planet, Ambasz seemed at once to burrow into the earth and to raise a confident structure in search of redemption. But most important, the house took its place squarely amid vibrant debates then under way in New York at a moment when the city, from the depths of a financial crisis, asserted itself as an intellectual outpost for the renewal of architectural culture and thought. Ambasz strode into the fray, adding a new architectural voice to the somewhat polarized conversation in 1970s New York, centered on the Institute for Architecture and Urban Studies, with which he was intimately connected from the conception of that now fabled architectural think tank, helping to draft even its founding documents, even as a few years later he would draw up a philosophical brief for a new kind of university—*Universitas*—in which design as a way of thinking could restructure the whole tree of knowledge and pursuit from the most practical to the most speculative.

Few at the time could realize his profound roots in the practice of Amancio Williams, then little known outside the Mar del Plata. For Williams, architecture was always as much a philosophical projection as an act of building. His diverse body of projects, many utopian speculations, presented always a new negotiation of the relationship between the buried and the floating, the territorial and the imaginary, most famously in the Casa del Puente (Bridge House) designed in 1943 for his father at Mar del Plata. Having worked for Williams as a teenager, these were images of architectural possibilities that Ambasz held on to as he encountered new positions in the intense debates of the 1960s emerging between Princeton, where he took two degrees in two years, and New York, where he aligned himself with both the Institute for Architecture and Urban Studies and the Museum of Modern Art, in the ambits animated by Peter Eisenman and the *éminence grise* of the postwar New York architectural debate, Philip Johnson. By 1975 the debate between the so-called Whites and the so-called Grays was reaching its climax, when Ambasz responded to the manifesto houses that had been brought together in a series of Museum of Modern Art publications beginning with Robert Venturi's *Complexity and Contradiction in Architecture* of 1965, which presented his fabled and gabled Mother's House, a great iconic facade behind which a complex set of design moves and historical references were brought into a taut but deliberately unresolved dialogue. Houses too dominated the portfolios of the architects brought together in 1972 in Arthur Drexler's *Five Architects,* in which the house became the matrix for new encounters between composition, or in the case of Peter Eisenman an apparent anticomposition, and philosophical hypotheses. Most potent for Ambasz it would seem, if we were to try to align his design for a House of Spiritual Retreat with the work produced in that widely influential white book, was the work of John Hejduk, and in particular Hejduk's first and second Wall houses. Between these two publications of the very department in which the young Ambasz was appointed curator of design in 1969 were gathered seemingly opposed quests for meanings, one treating architecture as a literary palette of charged symbols, the other as a generative grammar capable of creating new meaning from new forms. As would soon prove characteristic, Ambasz's design could be seen both as a response to either side of the temporary battlefronts and as a position that at once breached the conflict and yet staked out a completely personal, other territory. And it did so also by entering a lifelong debate with another figure with whom Ambasz was to have both real and imagined conversations, Philip Johnson, whose Glass House in New Canaan, the ultimate architect/historian/

curator's autobiographical statement, was equally a conversation with a spiritual mentor (in this case Mies van der Rohe) and with the history of architecture. It is only when these stakes are taken into account that one can begin to understand Ambasz's continual return to the extensive site of his estate outside Seville where he periodically adds new follies, new constructions, and new glosses on his own design, just as Johnson's New Canaan property—where the curator/architect slept in a closed volume in counterpoint to the famed transparent box—would serve as a picturesque landscape for the evolution of a design vocabulary and continual self-commentary. But whereas each of Johnson's projects marked a shift in his design language, material vocabulary, and changing force field of influences, Ambasz's work continues to develop and expand on the themes, both in dialogue with landscape and building, announced in the Casa itself. Ambasz's is a kind of allegorical quest for meaning, but one in which forms are often detached from specific referents.

To decode Ambasz's house with its myriad of associations and recollections held together by the improbable dualistic encounter of a monumental extruded corner of nonenclosing walls with a sunken living quarters tucked under a berm and facing into a sunken patio court, is to enter more into the world of Freud's *Interpretation of Dreams* than the art historical realm of iconography. Magical realism in South American literature probably has more to offer as a guide than the picturesque logic of the English landscape garden of the eighteenth and nineteenth centuries with their pavilions and ha-has which are no doubt distant cousins of Ambasz's work. But whereas these are generally archaeological *exercises de style,* Ambasz's designs bring forth a world of associations from the collective history of architectural imagery and the specialized world of architectural history, images both from shared dreams and those that might only interrupt the sleep of an architect steeped in the most extensive architectural culture. The monumental extruded "spirit" house, detached from its site as an illustration, could easily be juxtaposed with the widest variety of photosets, from the Andalusian houses of nearby Seville, Granada, or Cordoba (an early proposed site for the project before it had found an actual residence in the earth), to the work of Luis Barragán, whose imagery between poetry and vernacular Ambasz would introduce into the debate and the imaginary in 1976 in a highly remarked Museum of Modern Art exhibition—one of his last—to the Wall House projects of Hejduk, destined like Ambasz's to a long existence as a potent idea before being embodied in constructive reality. Indeed if Hejduk and his fellow Whites seemed largely to be deriving new sets of manipulations from the language of the European avant-gardes of the 1920s to create a neomodernist position, Ambasz set out to excavate a language from a much broader and deeper range of inspirations, ranging from Pre-Columbian spaces through the grotto imagery of the Renaissance to Claude-Nicolas Ledoux (notably in the fake rustic stone juxtaposed with smooth surfaces of the human hand). If the chain of references passes via Renaissance humanism and the French Enlightenment, these sources among many others are juxtaposed and merged in ways that have more to do with surrealism, in both its pictorial and literary practices, than with the more rigorous forms of collage that intrigued a number of other figures around New York's Institute for Architecture and Urban Studies.

Once inside the space defined by the great L-shaped wall, the theatricality of Ambasz's architectural vision is clear. Here the spiritual retreat offers a dramatic stage, with a new duality: a set of broad stairs descends toward the sunken patio onto which the principle rooms of the house face. Turning our view to the entry that has taken us, like Alice, through the looking glass, we discover on the inner face the walls sponsor a twin set of frighteningly narrow stairs anchored on one side into the wall, cantilevered out over the void at the other end. Rare are visitors who have the courage to mount the stairs to pass back through the wall, this time to a veranda suspended high over the landscape. Ambasz himself professes never to have been there. Here in the same years that James Turrell set about his first works framing the sky, Ambasz too created a frame for viewing the cycle of times of the day and changing lighting effects, as his rotated square plan of a house becomes a mechanism for contemplating extracted elements of the adjacent landscape, a stage where a drama is continually unfolding even absent actors and viewers, a theater of memory perhaps more than a memory palace. To traverse the two halves of the great square, one half descending majestically and slowly into the ground, the other offering a view of the sky flanked by "ladders" ascending to heaven in a way more inviting to the eye and the spirit than to the feet, is also to experience a series of reversals, most important, the oscillation between audience and stage, as though the hypotenuse that the sunken and elevated right-angle triangles share is an invisible proscenium, a site of continual exchange between stage and audience. And here for the first time we realize that Ambasz's work is not only about the duality between abstraction and symbol, but also the theatrical creation of a stage for the performance of rituals and ceremonies, suspended between a lost past and a yet-to-unfold future.

As the doors to the various guest rooms in the vaults of the subterranean house remain closed in the official photography of the house—provided by Michele Alassio for the lavish monographic documentation of the house published in 2005 to accompany an exhibit of the house at MoMA—we are left dreaming of the dialogues that might take place on that stage. Given the archetypal simplicity of the house's component parts we might project any number of scenarios that belong to the world of fables that so intrigues Ambasz. The architect or architectural historian will continually detect conversations between the absent designer, Ambasz, and a host of imagined guests: from the designers of ancient Greek theaters, via Bramante in his projecting of the Vatican Belvedere, a series of terraced gardens inside high walls, to Alvar Aalto in the play between columns and undulating walls in the buried house with a free plan set behind an echo of the Iberian patio, to name but a few that come to mind as we traverse the memory palace in photographs. If Robert Venturi had introduced the world of "both/and" in his early writings that were plumbed by all in the years Ambasz was speeding through Princeton, here we are in a world in which the stratigraphy of associations goes beyond the apparent duality the house sets up in the aerial view of its model.

For Ambasz's project in architecture is to be found as much in what doesn't meet the eye as in the intriguing reversal of the monumentally scaled house corner that turns out to be an L-shaped theater proscenium as in the ancient Roman theaters of antique Spain. For it is the nip and tuck strategy of embedding the spaces for living, dining, and sleeping in a berm that completes the raised knoll on which the house is set that was to provide a point of departure for Ambasz's development of an architecture which could seek to reconcile nature and culture in a world of advanced technology. Here Ambasz reveals himself to be as much a child of Earth Day, first celebrated in April 1970, as he was of the debating societies of Princeton and the IAUS in Manhattan. If we are to turn our technological archaeology to the layers of information embedded in the World Wide Web rather than in the layers of cultural residue of the landscape we can unearth, we are quickly led to the discovery that Ambasz himself has recently planted a clue. Invited to add his voice to the growing set of book lists on a website *Designers and Books,* Ambasz offers only five titles that have "all influenced me profoundly." He notes there "How can I extricate them from my memory? They are now substantially part of me."[1] After surveying a world of literary favorites from Aristophanes via Aeschylus, Sophocles, and Euripides, to Cervantes and Shakespeare, Ambasz reveals an influence one always feels in his contrarian spirit, his taste for the rationality of the absurd or the absurdity of rationality, Jorge Luis Borges, and he quotes J. M. Coetzee speaking of Borges: "He, more than anyone, renovated the language of fiction and thus opened the way to a remarkable generation of Spanish American novelists." Ambasz himself continues "a stylist of great fabulist imagination, superb elegance, and economy of means." Given his own claims for himself as the fabulist of architecture, this is an influence that demands to be taken seriously and explored in greater depth. This I must leave to a person versed in Borges's world, although it is clear that Ambasz too is after the construction of worlds, and not simply of buildings. But he leaves us two clues in the list of but five titles offered on a website where few contributors stop at fewer than fifteen. Not the least of these is Lucretius, as Ambasz explains:

> *On the Nature of Things (De Rerum Natura)* is a first-century BC epic poem by the Roman poet and philosopher Lucretius. It deals with the principles of atomism; the nature of the mind; explanations of sensation and thought; the development of the world and its phenomena; and explains a variety of celestial and terrestrial phenomena. The poem grandly proclaims the reality of our role in a universe that is ruled by chance, with no interference from gods. It is a statement of personal responsibility in a world in which everyone is driven by hungers and passions with which they were born and do not understand.[2]

This is a quote that is interesting to juxtapose with a philosophical position offered to clients and researchers on Ambasz's own website when he declares:

> We are beginning to understand that, like the ancient people of non-Greek cultures, we should see humanity not in contrast to, but as an integral part of both the natural and the man-made milieus. Man should not see himself as a separate entity, detached from nature, but should accept his existence as part of it. Similarly, the artifacts we create should not be proud aliens but rather should be designed as carefully and intricately woven extensions of the larger natural and man-made domains surrounding us.[3]

But to these erudite references, Ambasz adds two unexpected masters of architecture, Le Corbusier and Frank Lloyd Wright, figures long represented as in a dialogue that for many is irreconcilable but which might also illuminate something of the juxtaposition of opposites in the monumental presences and buried absences in Ambasz's architecture. In addition to Henry-Russell Hitchcock's monograph on the American master, published by the Museum of Modern Art in 1941, Ambasz offers Wright's own *The Natural House*, first published in 1954. One has the sense then that Ambasz's own project, ever since his 1975 house design, has been to effect a dialogue not only between the machine imagery of Le Corbusier and the natural imagery of Wright, but between the philosophical projects of the Roman philosopher Lucretius and the organic strain in modernism. As Ambasz recalls touchingly of a book that shaped his views just as he was on the verge of encountering the work of Amancio Williams: "This book is one of several Wright wrote to proselytize for his notion of organic architecture. I read it when I was fourteen years old. Stylistically abominable, it is nevertheless a very influential text. Organic architecture is a philosophy of architecture that promotes harmony between human habitation and the natural world through design approaches so sympathetic and well integrated with the site that buildings, furnishings, and surroundings become part of a unified, interrelated composition." From the first this was no mere formal challenge but a project that could variously engage a political potential, one perhaps clearer in Ambasz's earlier work than in some of his most recent structures.

And it is with the example of Wright that we might begin to think about Ambasz's strategy of layering a building into the earth, with a combination of incisions and berms, in a set of techniques where the practicality of thermal protection—either from the extreme heat of the south of Spain or the often severe cold of the Northeast of the United States—meets the poetry of an architecture of landworks. Here Ambasz conjugates meanings, references, and techniques that take his historically from-the-earth works of earlier civilizations to the half-buried architecture of Wright, most famously in the second Jacobs House of 1944 where from one side the house appears to merge with the earth. Here he demonstrates that a ritual, for what Fulvio Irace has aptly called "a technological Arcadia," could be developed as much from a monumental ceremonial of landforms as one

Cooperative of Mexican-American Grape Growers, California, USA, 1976

of built symbols. Rural and urban appear at first to stand in conflict. In the countryside Ambasz likes to tuck and fold, gently inserting his building into a landscape in such a way that they generally seem at once in harmony with their natural setting and to subtly reorder it, almost like the ancient architects of Monte Albán outside Oaxaca, where monumental engineering seems but a working with the found landscape. From the first this strategy is more than ecological, engaging also literally or poetically, a social project. Nowhere was this more dramatic than in one of his early works—and critics rightly underscore the extent to which the clarity of a social position has become increasingly obscured as the techniques of Ambasz's architecture have become transferable from project to project, country to country—the 1976 project for a Cooperative of Mexican-American Grape Growers designed at the height of César Chávez's campaign for the rights of farm workers and particularly of the Latino community in the American Southwest, many of whom were involved in the grape-growing industry of Southern California. Here, in a rare literal engagement with current events, Ambasz created a landscape in which a much more humane vision of the backbreaking labor of cultivating grapes derived a poetic civic agrarianism out of a solution to both the practical and spiritual problems of a huge community of Mexican Americans, all of it imagined for a site near a border that today is a site of conflict and the violent encounters of two different economic regimes. A troubling social reality became the raw material for reconfiguring a landscape productive of a better climate and better social relations, of diginity and poetry, even from the stark realities of modern emigrant agriculture life. Like so much of Ambasz's work it seems in retrospect remarkably anticipatory of the dilemmas of the twenty-first century.

But it is in his intervention in urban landscapes that Ambasz offers a third way, creating an artificial nature for daily urban life with no pretenses to restoring ecology as found. In the late 1970s Ambasz had declared already—decades before the strategies of current landscape urbanism—that an architect's tools, and his palette, could extend beyond traditional building materials to plants, to water, and to light. These themes had been approached somewhat formulaically in the work of Kevin Roche and John Dinkeloo, notably in the Ford Foundation completed in 1968, and in the series of hotel interiors developed by Atlanta architect John Portman in those same years. But Ambasz was afraid neither to seize their techniques nor to develop a type of populist model for his renderings that seemed almost designed to offend the taste culture of his fellow intellectual architects in the New York of nascent postmodernism. But out of these strategies that Portman in particular had used for added commercial value, Ambasz began to craft a new vernacular of a landscape civic life that embraced artificiality, even as it harnessed the environmental and thermal properties of plants in the crafting not of scenery but of environments. While New York was debating the relevance of the Beaux-Arts tradition provocatively displayed by Arthur Drexler in a landmark, but for many enigmatic, show of grand drawings at the Museum of Modern Art in 1975, a display which coincided with the battle to preserve such American masterpieces of civic monuments in the Beaux-Arts style, notably Grand Central Station in New York, Ambasz proposed a revitalization of a Beaux-Arts building for the Grand Rapids, Michigan, Museum of Art, in which the building would literally be invaded by a huge greenhouse. Ambasz took the types of glass conservatories that nineteenth-century French architects often hid behind their perfectly ordered facades and transposed it to the public front of the revitalized museum to become a major motif, an almost surreal other to the masonry frame. This was a strategy developed further in two seminal projects of the 1980s: the 1982 project for recasting Salamanca's Plaza Mayor by the excavation of a multistory grotto at its center, and the 1986 project for revitalizing Kansas City's 1914 Union Station. Just as the ethos of historic preservation was taking shape in which new commercial activities could reanimate partly abandoned civic structures in troubled American downtowns, Ambasz put forth the provocative gesture of grafting nature and culture, even framing the environmental benefits of bringing greenery into cities. Not surprisingly, these projects must be read historically as critical of the emerging commercial paradigms, so it must be recognized that they contain ideas with an astounding early twenty-first-century actuality.

Plaza Mayor, Salamanca, Spain, 1982

Union Station, Kansas City, Missouri, USA, 1986

Nowhere is this prescience more striking than in one of Ambasz's finest designs, his 1975 proposal for a Center for Applied Computer Research for Mexico City. Not only did this project accurately predict the extent to which in a short period of years the computer would radically change the ways in which we inhabit the land, structure our days, and relate to one another, it also provided a model for the symbiotic ways in which advanced technology and ecology could leverage the capacities of each to restore a place for man in a heavily damaged planet. As Ambasz himself later described the project, "Behind the design of this environment is the premise that nobody should have to work. At worst, one would work at home and not need a large building but rather a small one to simply house a computer and receive messages. The building has been conceived, therefore, as a set of elements that can be progressively reconfigured and recombined as the needs of the office vary over time."[4] Not only does this prediction correspond precisely to the ways in which the computer is reordering space and social relations at every scale from the desk to the city, but the radical proposal to create the whole as a series of floating units in a restored lake landscape of Mexico City predicted, nearly twenty years before it became a pressing concern, the need to restore Mexico City's long-lost ecology not out of landscape nostalgia but out of vital necessity to keep the city from sinking due to the gradual undermining of its foundations and to reintroduce the vital humidity of the natural aqueous ecology. In 2011 London's AA staged a summer school in the Mexican capital called "Recovering Waterscapes," focused on the type of urbanism that would result from the restoration of a long-lost natural resource in a country with ever greater

Center for Applied Computer Research, Mexico City, Mexico, 1975

catastrophic water shortages. If I have dwelt at length on a series of projects from the first five years of Ambasz's architectural career—the Casa de Retiro Espiritual, the project for the Grand Rapids Art Museum, and the Mexico City office landscape—it is because in a very short period of time with these projects Ambasz crafted a typology of concerns that would define a project in architecture that has continued with remarkable consistency for more than four decades since. Little of his vast catalog of work has been realized, but each new project explores the further applicability of a set of inventions—for in the end as architect as well as industrial designer Ambasz is first and foremost an inventor in which the most improbable seems ripe to merit a patent, like the modeling of a chair back on a vertebra. If his architectural models, beginning with a house for Barbie, embrace a self-conscious populism bordering on kitsch, the deliberate declaration of artifice in fact carries over into the built work, which in no way claims to restore something that was once whole and is now lost, be it in landscape or in the increasingly prominent visions of a lost small-town life that have permeated everything from the imagery of the shopping mall to New Urbanist projects for crafting an urbanity of the future. Ambasz refuses such nostalgia and impossible wholeness even as he crafts populist imagery. Recent built projects like the Banca degli Occhi in Mestre of 2009, no less than the remarkable Fukuoka Prefectural International Hall in Japan of two decades earlier, bring gardens into dense landscapes of modern commercial reality without any attempt to suggest the restoration of a lost Arcadia. Rather then revel in the very artificiality of terraces, trees planted in boxes and piled in grids, or stepped

Grand Rapids Art Museum, Grand Rapids, Michigan, USA, 1983

facades of greenery are more reminiscent of the ancient tales of Babylon or the fabled dreams of Kubla Khan.

As much as Ambasz has developed over decades a poetry of architecture, he was also prescient in exploring a way in which the techniques of an environmental architecture could dovetail with an architecture that supported a renewal of ceremonies and aura, of ritual and signification, however elusive. In a world in which buildings are awarded numeric ratings on environmental accomplishments, so-called LEED ratings, Ambasz's collected work is a powerful reminder that a meaningful architecture of ecological inhabitation of an increasingly imperiled planet is not a challenge for technology alone, nor for practicality. It is often the blatant improbability—in which architecture and nature seem at once reminiscent of something lost and projective of something never before seen—which brings new visions for the possible.

1 www.designersandbooks.com/designer/booklist/emilio-ambasz
2 Ibid.
3 www.emilioambaszandassociates.com
4 Emilio Ambasz et al., *Architecture & Nature/Design & Artifice,* new rev. ed. (Milan: Electa, 2010), xliv–xlv.

La Casa de Retiro Espiritual

Seville, Spain, 1975

Situated outside of Seville, in the middle of a rolling wheat field, this is the weekend retreat of a couple with two children. Inspired by the traditional Andalusian house, it has a central patio onto which all rooms open. The house is insulated by the earth which keeps it naturally cool in the hot, arid southern climate. Two tall, white, rough stuccoed walls meet at a right angle to herald the entrance of the house which is otherwise contained by sinuous walls. On both the high walls water cascades along the handrails, creating a significant amount of noise when one is at the bottom of the stairs but becoming increasingly quieter as one ascends. The large, continuous living space of the interior area is defined by smooth cavities excavated in the floor, which are echoed in the ceiling above. Small, inlaid glass tiles are washed by the soft, diffused light that falls from the skylights and filters in from the breezy patio.

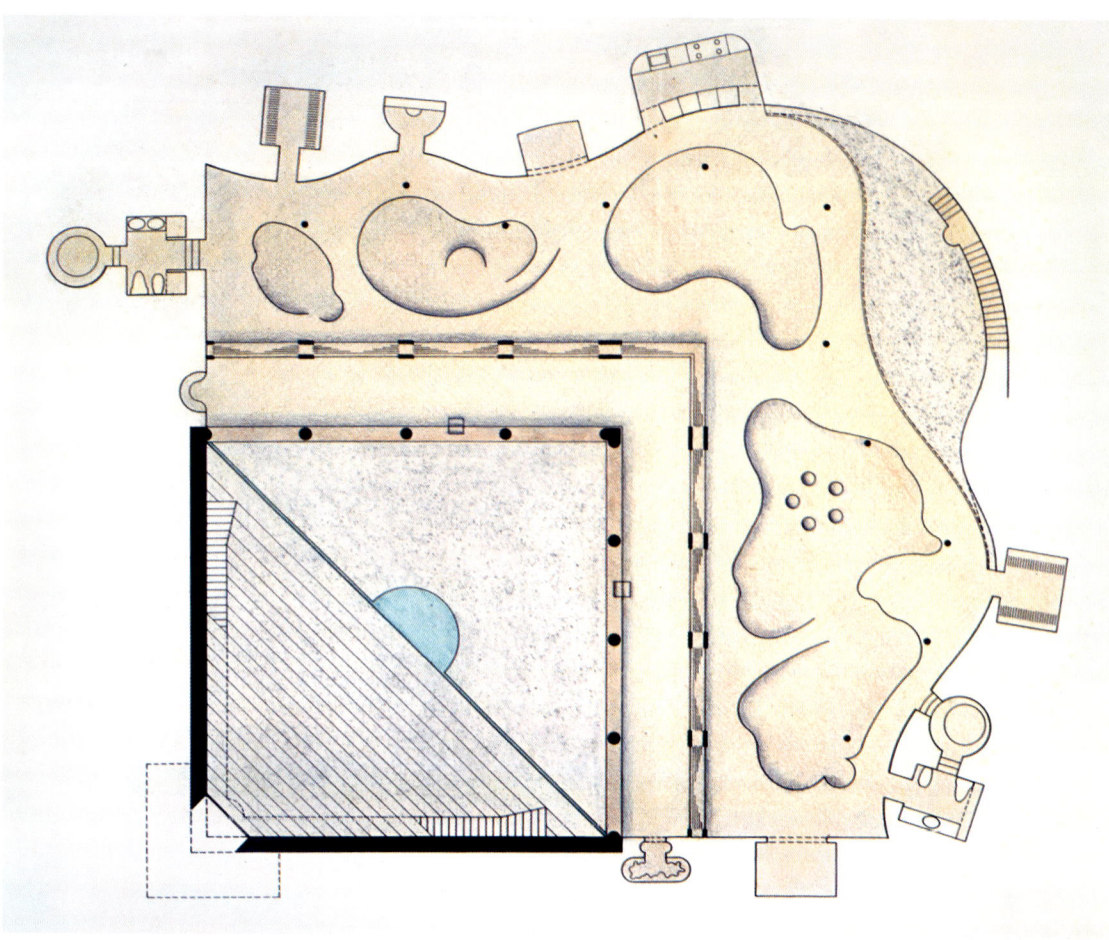
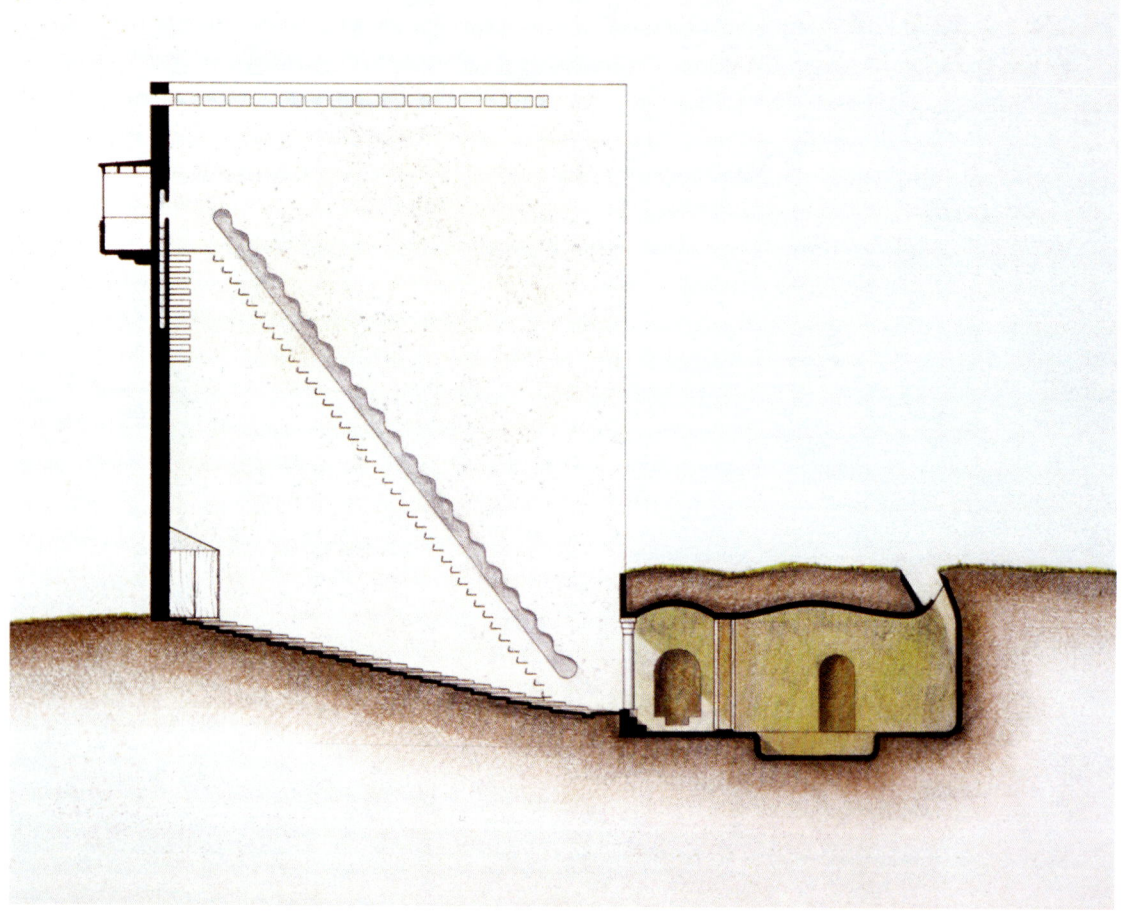

Ambasz's work offers no generally applicable panaceas; but is at least one such provocation to ponder the great cultural project of the near future, one necessary to ensure we might have a future.

Peter Buchanan

THE ARCHITECTURE OF EMILIO AMBASZ

Peter Buchanan

The ideal gesture would be to arrive at a plot of land which is so immensely fertile and welcoming that, slowly, the land would assume a shape—providing us with an abode. And within this abode—being such a magic space—it would never rain, nor would there ever be inclemencies of any other sort. We must build our house on Earth only because we are not welcomed on the land. Every act of construction is a defiance of nature. In a perfect nature, we would not need houses.[1]

Emilio Ambasz

Looking over the architectural oeuvre, built and unbuilt, of Emilio Ambasz, the most immediately obvious characteristic is a tendency to blur the distinction between architecture and landscape. Typically, it is difficult to say where one begins and the other ends. Instead of buildings standing proud from the surrounding, often subordinate, landscaping, the latter tends to envelope and sometimes even penetrates Ambasz's architecture. In most cases earth and vegetation sweep up and over the building, which typically is built above the ground with earth banked up against and over the top of it. And the building's presence is enigmatically announced or shyly revealed by isolated elements projecting through and above the landscape. These may be prominent freestanding walls, a curving colonnade bulging forward through an earth bank, wiggly retaining walls, or tall glazed roofs of different geometric configuration. They also may be sunken courts or low skylights only discovered when you are almost on top of them. In many other schemes each floor steps back from that below, creating a series of plant-covered terraces, so becoming a verdant hill or peak within the surrounding landscape or an extension of a public park. On yet other buildings, those that rise tall above the ground, plants swarm over the elevations to festoon, shade, and soften them.

What is this all about? And why do many find it all so appealing and even topical? An obvious reason is that people see this as a vision of a "green" architecture that increasing numbers understand that we will have to embrace, and some already long to do so. The blurring of building and landscaping suggests to them some reconciliation between culture and nature, thus reversing the several-millennia-old quest to subjugate a hostile nature and instead seek a symbiosis with a benevolent nature on which we have foolishly forgotten our dependence. Somewhat more prosaically, people increasingly regret the loss of countryside and its once varied flora and fauna, as well as the diminished presence of greenery in so many contemporary cities. Ambasz's motto of "green over gray," implying vegetation swathing construction and the protection and expansion of greenery, thus has immediate emotional appeal.

Another, related, reason for the appeal of Ambasz's work is probably because people respond, perhaps only semiconsciously, to a poetic vision that seems to inspire the designs and that speaks to something deep within us. The works hint at, or suggest they might take us back to, some state of prelapsarian innocence, a time—whether in a mythical past or childhood experience—when the world was still enchanted. The designs seem born of a dreamlike vision of a benignly hospitable Earth that heaves open and invites us to shelter within its embrace as soil and plants draw up over us as a comforting, protective blanket. We feel we can nestle safe and sound within this benevolent embrace, rather like a child snuggling beneath the bedclothes as the wind howls and the rain drums on windows and roof. But even more than that, when approaching and within this semisubterranean realm, we are presented with a choreographed sequence of experiences that not only calm our contemporary restlessness but also open us up to connect with things we are too often oblivious of, such as the sky and stars above and our own deeper selves. Unsurprisingly, the most enchanting of these designs arise from reverie.

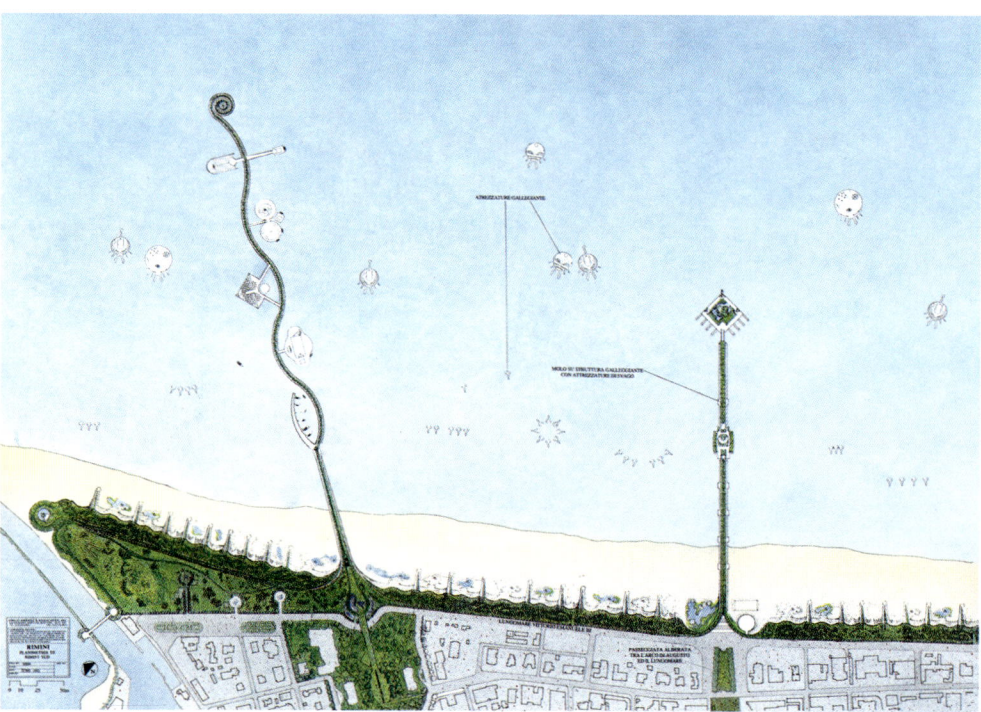

Rimini Seaside Development Center, Rimini, Italy, 1990

Ambasz reports that he designs by allowing images to arise from his imagination unsullied by verbal description and interpretation. Only later, once the images have elaborated and clarified themselves, does he begin to comprehend how they might host human life and its daily rituals; and only later still will he attempt intellectual interpretation. For Ambasz, words are inevitably connected to what one knows, and so to the past, and it is only by giving primacy to the unmediated image that the truly new and unexpected arises.

Now that Ambasz has amassed a considerable body of architectural works and is building more and more, it is possible to see each of his works as placed somewhere on a continuum between two poles. At one extreme are the sort of schemes already alluded to, those that seem most likely to have emerged from reverie and which, no matter how pragmatically realizable, have an irrational—or is it suprarational?—dimension from which also arises much of their potent poetry. At the opposite pole are the entirely rational and eminently sensible schemes. This is no implied put-down: these schemes are exactly apt as solutions for very real problems of general relevance and deserve to be widely emulated—unlike those of the contrasting pole for which that would probably prove problematic for lesser talents.

Typically the schemes of this pragmatic, professional pole introduce (or reintroduce) verdant vegetation into a city center or between it and a beach. In these designs, as with virtually all of Ambasz's work, the vegetation or landscaping serves a sheltering function, as with the barrel-vaulted pergola leading to Frankfurt's Eschenheim Tower (1985). Or, with the Rimini Seaside Development (1990), where a proposal to upgrade the resort with a slightly raised, verdant and shady linear park between city and beach—which contains year-round attractions such as bars, cafés, and picturesque walks and would also ameliorate the local microclimate—shelters beach-related facilities at that level. The master plan for Barletta (1997) is a more complex version of such a scheme, also linking city and sea but overcoming a wider range of obstacles on the site to accommodate a more varied functional program.

A particularly engaging and ingenious urban landscaping scheme is that proposed for the Frankfurt Zoo (1986). Here each species would occupy its appropriate habitat within what seems a continuous but varying landscape with animals contained and separated from people not by fences but by apparently natural features such as pools, rock faces, and other changes in level. The climactic focus of the scheme appears to be the remnants of a small extinct volcano and in this tall steep-sided crater would be creatures that fly and swing in the tree canopy. Another ingenious, somewhat crater-like scheme bringing greenery into a congested city center, as well as extra housing and other facilities (which because under planted terracing do not diminish the surface area of new landscaping), is the proposal to dam and drain the inland portion of the harbor basin at the Old Port of Monte Carlo (1998). Ambasz's first realized design bringing more greenery into a city center is the Fukuoka Prefectural International Hall (1990), where an existing park is extended as a garden climbing in terraced steps up the full height of the tall building. The distinctiveness and ingenuity of all these schemes distance them somewhat from the extreme pole of more pragmatic professionalism represented by the beachfront proposals.

Frankfurt Zoo, Frankfurt, Germany, 1986

But to return to the opposite pole mentioned first, the schemes that seem most directly derived from dreamlike reverie, with somewhat startlingly poetic consequences. The most enchanting and memorable of these are organized around a processional route so that they would be engaged with over time in a ritualistically structured sequence of experiences. This pole is exemplified by a number of houses, a pair of plazas, and even the way entrance is handled into the American Folk Art Museum tower (1979). The architectural design that first brought Ambasz to international attention in 1979 is what would later be realized as the Casa de Retiro Espiritual near Seville, discussed at length elsewhere in this catalog and so not described here. It is now to be joined by a private architectural museum. Set some distance away, this too will be bermed over with earth and planting from which will rise a solitary tall wall that signals some relationship with the house. Against this wall, a portion of the earth heaves up at an undulating angle to invite you down into the museum.

Designed in a somewhat similar vein, but along a more protracted processional route, is the house for Leo Castelli in East Hampton, Long Island (1980), a shyly recessive scheme despite its graphically potent geometric composition when seen in an overhead model view and the grandeur of the entrance portal to the grounds. The latter is

Public park and residences, Monte Carlo, Monaco, 1998

defined by a pair of massive earth berms that mark and frame the beginning of the processional approach, as did the pylons before ancient Egyptian temples. The presence of these mighty earthworks and the upward slope of the path between them would inevitably slow those approaching until they found themselves raised enough to pause before a panoramic view and suddenly discover they were above the corner of a sunken court set at a diagonal to the approach. From here a widening flight of stairs descends to where the lowest step bisects the bottom of the court. As at the Casa de Retiro Espiritual the effect will be to slow both your rate of approach and your sense of time so you reach the semicircular pool marking the center of the court both tranquil and alert. Edging the two far sides of the court is a colonnaded ambulatory overlooked by bedrooms. Beyond the corner of the ambulatory is the living room that bulges forward from the enclosing earth to command the broad view ahead, the centrifugal and horizontal extroversion of the room contrasting with the centripetal introversion of the still court with its implied vertical *axis mundi* marked by the central pool.

Admirable here is how abstraction and understatement can be used to such experientially rich and potent effect. If the Casa de Retiro seems best suited to solitary life, or at most a couple, this house seems suited to a family and guests; again, though, the entrance sequence would be best experienced alone when undistracted by company and conversation. And marvelous and moving as this protracted processional—with its chthonic, sacred overtones—would be to experience, is this perhaps not too compelling a ritual to be indulged in on an everyday basis?

Casa Canales, Monterrey, Mexico, 1991

A house with a potent entrance sequence that would work well if experienced in a group rather than only alone is the Casa Canales designed for a mountainside site perched above Monterrey, Mexico. This too offers a panoramic view just prior to descending to enter it as well as a corner living room with a broad extrovert view. But otherwise it is very different from the two previous houses. Here the car is overtly acknowledged with space for several to park behind and above the house, under a pergola that shades those arriving from the fierce Mexican sun. Extending forward from the pergola, reflecting the sky and surrounding mountains as well as insulating the roof of the house below, is a large triangular pool filled with water from an abundant spring above the house. Along the outer edge of the pool a walkway leads to a pavilion at the apex of the triangle. Like the *chhatri* in Mughal architecture this is a calmly contemplative place, defined and dignified by the peaked roof that asserts a scale suited only to a single person or small group, from which to enjoy views and breezes, especially in the cool of mornings and evenings.

Entrance to the house is by a broad stair that descends through a slot in the pool to eventually arrive two levels down at the lofty living room overlooking a colonnaded veranda and the view beyond. Other than at the corner, the double-height veranda is shaded by a leaning lattice supporting fragrant flowering plants and penetrated by large window-like openings to give views to the rooms behind — various service rooms on the living room level and the bedrooms on the floor above. The generous scale of the entrance pergola, broad pool and tall veranda, and the lattice frame with its arched window openings and lushly fragrant infill, all result in a house with none of the austerity of the former pair, and lacking the calm central court it is entirely extroverted rather than introverted. Yet it is no less enchantingly poetic. If the other two houses, like Hispanic architecture generally, draw on Arabic roots, then this one seems altogether more Indian in inspiration. Some might see it as more feminine than the somewhat austerely abstract courtyard houses.

Also for a hot climate, which is humid too, is the Houston Center Plaza (1982), designed to suggest a center within the city's extensive urban grid and for cool, shady relief from the sun and heat. From the sidewalk edge, the floor of the plaza dishes downward to a pool raised again up to sidewalk level, from which water cascades around its perimeter as well as inward to some hidden center from which rise swirling mists on which dance colored laser beams. Between sidewalk and pool is a grid, a microcosm of the larger urban grid, of fragrant, flowering vines supported on steel frames and mesh, their tops always level with the sidewalk and pool. Where these become tall enough, openings in their sides permit entry into these bowers, scaled to accommodate intimate trysts, quiet reading, games of chess and so on. Pedestrians passing are drawn inexorably inward by the sloping floor, the cooling mists emanating from the top of each plant-swathed frame and the mirage-like vision of the central pool and its mist-generating cascades. Getting closer to the corners of the pool, gaps appear in the curtains of water allowing entry to a ramp that spirals downward, behind cascades stepping inwardly toward the center, to reach a variety of cultural facilities and eating places.

By contrast, the proposal for the Plaza Mayor in Salamanca, Spain (1982), is to transform an existing historic plaza, also to accommodate subterranean cultural facilities. Here too the plaza dishes inward, using rows of steps and landings rather than a steadily ramping floor, the whole shaded by the spreading canopy of trees rising through the plaza floor from planters all at the same level. At the floor level of the existing surrounding arcades would now be a sea of soft green foliage. Into this you'd enter and descend as if a swimmer into surf to discover below a quiet and enchanting, shady semi-underworld suited to all sorts of solitary and communal activities — below which is a real underworld of cultural facilities. It is a wonderfully appealing vision, and would be a beautiful place, but is quite unsuited to Latin culture. The present plaza works perfectly for the evening *passagiata* and to sit at the café tables around the perimeter enjoying coffee, beer, or a *fino* while watching those passing.

Close to this pole, not least in also being organized around processional routes, are the various garden schemes, particularly those that include gardens set below the new, raised ground level and capped by tall conservatory roofs. This series starts with Ambasz's first executed architectural work, the Lucile Halsell Conservatory in San Antonio, Texas (1982). It also includes the Baron Edmond de Rothschild Memorial Museum for Ramat Hanadiv in Israel (1993) and the Thermal Gardens (1996) for Sirmione, a spa on Lake Garda. This potentially delightful scheme would have coped with the demand for increased visitor numbers without loss of landscape amenity. Perhaps somewhere in the middle between the poles are museum designs

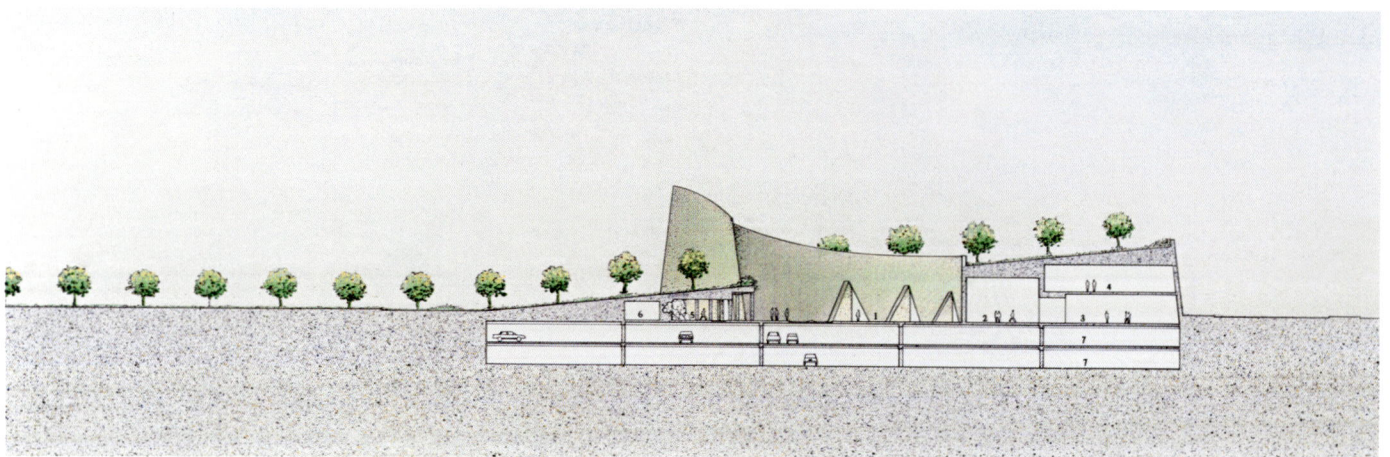

Phoenix History Museum, Phoenix, Arizona, USA, 1989

that are wholly or partially covered in earth and include courtyards. These too are arranged around a processional route, although one lacking (quite aptly) the intensity of those of the houses and plazas described earlier. Particular examples are the Phoenix History Museum for Phoenix, Arizona (1989), where a tall curving wall rises to announce its presence and define the entrance, and the Glory Art Museum for Hsin-Chu, Taiwan (1998), where the bulk of the building is exposed but the landscaping rises up the gently sloping roof to be used as a sculpture garden.

Also somewhere midway between the polar extremes are the various projects in which the roofs step progressively back at each floor up and the exposed roof is planted giving the impression of a terraced hill or peak. These schemes, all designed to impinge minimally on the surrounding landscape, include the Worldbridge Trade and Investment Center in Baltimore, Maryland (1989), the Complejo de Oficinas for La Venta, Mexico (1993), and the Winnisook Lodge in New York state's Catskill Mountains (2000). A somewhat similar strategy of planted roofs and terraces, although here also including sloping mesh to support screens of climbing plants is used on a large scale at the Nuova Concordia Resort near Castellaneta, Italy, now completed. It was this strategy that brought permission to build on land alongside the protected forest that extends along a pristine stretch of the Ionian Sea. Indeed, many of Ambasz's current commissions come his way because his approach gains acceptance where other architects have failed.

Ambasz has also designed schemes that are much too tall for the strategies described so far to be practicable. These include the ENI Headquarters for Rome and a Commercial and Residential

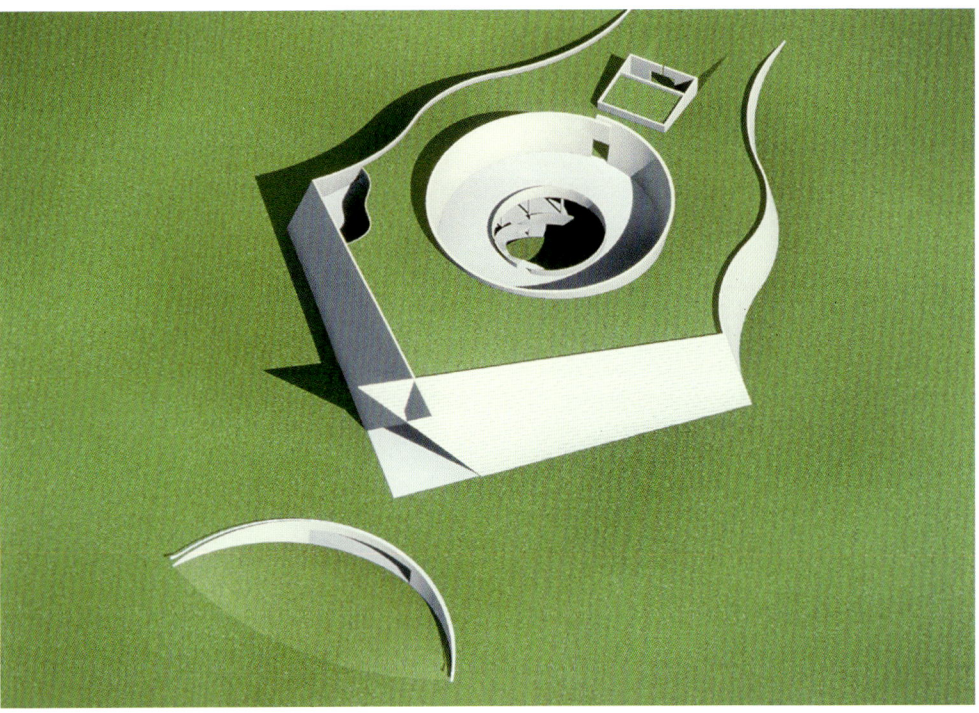

Glory Art Museum, Hsin-Chu, Taiwan, 1998

Development for The Hague, the Netherlands (2002). Here he has bermed over parts of the lower levels, but otherwise extended ledges, balconies and frames beyond the external skin of the building to support plants of various sorts, a strategy he has also used on lower buildings such as the Ospedale dell'Angelo (2008) and the nearby Banca degli Occhi (Eye Bank) (2009) in Mestre, Italy. In the first of these a sloping glazed roof—a device proposed for several projects and first realized at the Mycal Cultural and Athletic Center in Shin-Sanda, Japan (1990)—encloses a large conservatory so that the wards looking in on it, like those on the other side of the block with planted balconies, both enjoy verdant outlooks. This is only one of several examples where landscaping penetrates the building.

In comparison with the hospital, the Eye Bank gains in potency as a composition from being more compact and from its triangular configuration. On the two shorter sides of the isosceles triangle are rhomboid, copper-clad walls whose pointed ends stop just short of meeting above the opening to the entrance court—a gesture of dramatic tension inspired, Ambasz reports, by the way the fingers of God and Adam reach to each other on Michelangelo's Sistine Ceiling. This gesture marks the beginning of a relatively short processional route that axially bisects the building to end in a sunken circular court half a level down, a portion of which is enclosed as a reception room and foyer to the lecture theater under the entrance.

The ongoing Piazza della Visitazione project for central Matera, Italy (begun in 2009), a complex project to accommodate cultural, commercial, and leisure facilities along with trains, a station, and car parking—thus becoming a new entrance gateway that enhances

Piazza della Visitazione, Matera, Italy, 2009

the identity of the city — reaches a new level of synthesis in Ambasz's work. Bringing abundant greenery into the center of the city and connecting up its parts with verdant footbridges, it clearly relates to the more purely professional pole of Ambasz's work. Yet it also combines these with formally assertive and allusive elements from the other, poetic pole, such as the piazza framed by three immensely tall walls and the canyon seemingly scoured by the watercourse along its bottom.

Also new here, though presaged by the interior for the Banque Bruxelles Lambert in Lausanne, Switzerland (1981) and the overly literal forms of the Monument Tower Offices for Phoenix, Arizona (1998), is that these elements allude to features found in the surrounding countryside. Thus the piazza enclosed by the three tall walls — each faced with a lattice concealing stage lighting, speakers and so on for hosting various forms of live performance such as concerts, theater, and opera — deliberately recalls a huge nearby quarry that was in the past put to similar uses. This same space is also the spectacular outdoor foyer for a more conventional theater set below a pergola-shaded garden and entered through a horizontal slot at the base of the end wall of the piazza. And the canyon is both a shopping center and gives access to the train station while also echoing the canyons that are a feature of the mountainsides around. (Historically, Matera is famous for its extensive areas of cave dwellings on hill and canyon sides.)

The proposals would thus not only improve the city as a place to live, with an expanded and memorable civic image, but also bring alive connections across time and space to Matera's history and rural

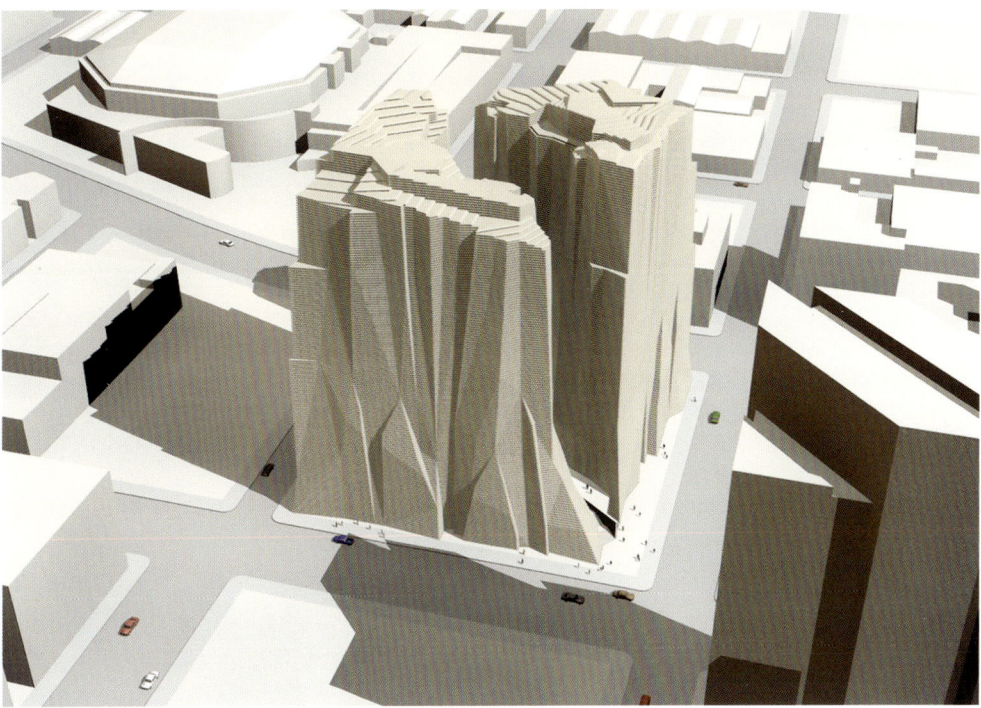

Monument Tower Offices, Phoenix, Arizona, USA, 1998

surroundings. Such a quest for reintegration and reconnection, as well as for enhancing urban life, is intrinsic to the larger quest for sustainability as we try and heal the fragmentation of our cites and the degradation of the environment wrought by the modern era.

Although Ambasz's influence is evident in some designs by other architects, which are not only partially earthed over but have features such as entrances strongly reminiscent of those in one of his projects, his architecture is quite different from that of anybody else. His work clearly belongs to the green camp, the liberal use of earth berming providing thermal insulation and thermal inertia, so ensuring comfort and diminished energy consumption. And the conservation of greenery and extending it through and alongside dense city fabric offsets the "urban heat island effect" that now makes many cities in hot places almost unbearable. But even among those also pursuing green designs, his work is distinct.

Virtually all of these other architects approach green or sustainable design as a technical and ecological issue to be approached scientifically and objectively. This it must be, and we are learning much from this work and its increasingly improved performance and broadening range of concerns and solutions. Yet also needed is engagement with subjective issues that extend beyond quality of life (rather than standard of living) to include psychological, cultural, and even spiritual factors such as awe and reverence for nature. More than technical issues and demands to cut consumption and emissions, these excite people and inspire them to commit to change.

We are at a pivotal point in history when we need to reappraise the very purpose of many things. Consider the difference implied in the names between agribusiness, a purely exploitive concern, and agriculture, which includes culture, so implying a whole way of husbanding Earth and its resources to ensure the long-term well-being of both it and ourselves. Something similar applies to architecture which should not just be about the provision of shelter, function, and economic return on investment. The wellsprings of architecture are in ritual as much as in providing shelter, in the spatial mapping and choreography of experience so as to elaborate and intensify it and enhance our sense of connection with our deep selves, the world of nature and even the cosmos. This is part of how we have evolved our cultures and have developed our sensibilities to evolve ourselves. Although modernity brought very many benefits, it and most modern architecture tended to downplay and even deny such concerns—which is one reason both have proved utterly unsustainable. We now need inspirational pointers to a very different future. Ambasz's work offers no generally applicable panaceas; but is at least one such provocation to ponder the great cultural project of the near future, one necessary to ensure we might have a future.

1 "Fragments from My Credo," in *Emilio Ambasz Inventions: The Reality of the Ideal* (New York: Rizzoli, 1992).

I've never been so much a man of leanings as much as I have always been a man concerned with going forward. In my work I have always striven to present alternative models of the future so we can change the present. This is a task to which I have dedicated my heart and mind. EA

Lucile Halsell Conservatory, San Antonio Botanical Garden

San Antonio, Texas, USA, 1982

The Lucile Halsell Conservatory is a complex of greenhouses located in the hot, dry climate of southern Texas. Unlike northern climates, where traditionally glazed greenhouses maximize sunlight, the climate of San Antonio requires that plants be shielded from the sun. Emilio Ambasz's design uses the earth as both a container and a protector of the plants, controlling light and heat levels by limiting glazed areas to the roof alone. This innovative design concept significantly decreased the need for expensive mechanical systems, thereby reducing the overall cost of the building by more than 20 percent.

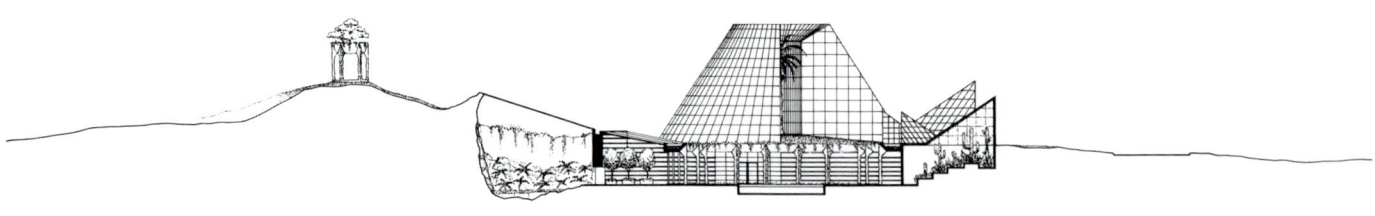

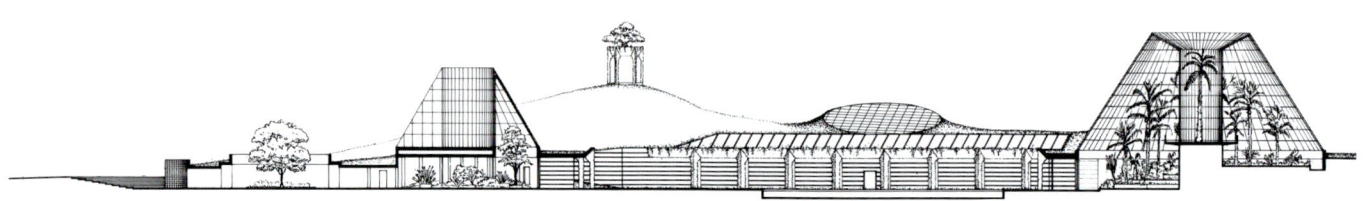

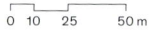

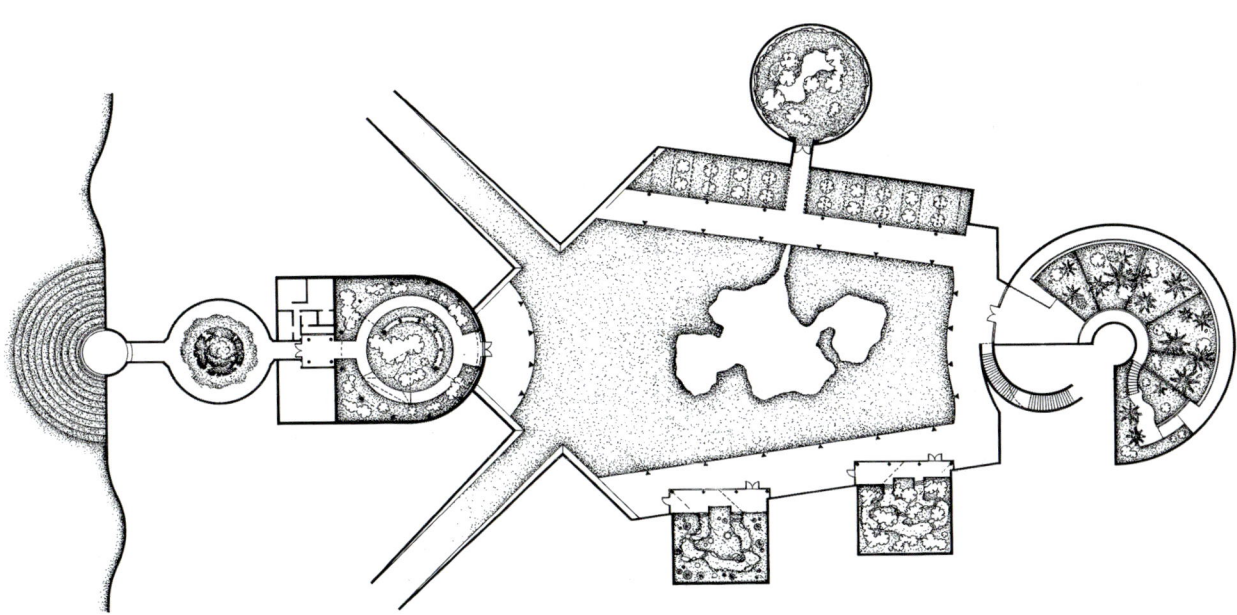
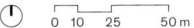

To state the issue baldly: Ambasz has brought the house down to the ground.

Lauren Sedofsky

PERIPHERAL VISION

Lauren Sedofsky

Given the sheer quantity of attention that has been directed over the past two decades toward the limits, the margins, the peripheries of artistic and intellectual practices—their repressions, suppressions, and proscriptions—it would seem only logical that Emilio Ambasz's work should have been picked up and picked over long ago as a unique effort to carry architecture across one of its most forbidding lines of demarcation into landscape. Yet, to date, even the seemingly inevitable observation that Ambasz has engaged in a conspicuous inversion of the respective weights conventionally assigned to the architectonic and the natural terrain has never been made, much less examined, with the clarity that it fully requires. The question is why, at a particular historical moment, this line of demarcation would have elicited Ambasz's disruptive investigations, while at the same time obscuring their implications for the potential power of landscape to intrude upon the autonomy and legitimacy of a bona fide architectural object. To state the issue baldly: Ambasz has brought the house down to the ground. And if familiar critical boundaries are fortified against such a move, it may be time to explore the unfamiliar.

 As far back as 1980 Manfredo Tafuri was to count Ambasz among architecture's young ironists.[1] By irony he meant an "interrupted criticism,"[2] the intermittent allusiveness of a poetic image. If the illustrations in *The Sphere and the Labyrinth* (1987) are any indication, Tafuri was thinking of Ambasz's proposal for the Grand Rapids Art Museum, involving the insertion of a monumental inclined stepped-plane cascade between the Beaux-Arts wings of an unused federal building. Right on target, Tafuri identified this massive intrusion of reflected sky into the building's academic vocabulary as an effort, typical of the Argentinian contingent in New York, to achieve a "masked architecture," which he equated with a "dissimulated language."[3] But what Tafuri ignored in Ambasz's "too-simple instruments" was their illicit foreignness. What he misjudged, out of a commitment to the totalizing power of architectural language, was the extent to which the masking gesture might constitute a historically determined statement. For, in the intervening years, Ambasz has developed just this gesture into a full-fledged, relentless principle, one that today, in its deliberately ex-centric way, would satisfy Tafuri's criterion for a genuinely critical architecture: "the systematicity of the heresy."[4]

The twist in architectural logic that has become Ambasz's modus operandi is hardly removed from the pivotal debates of the moment. The furious oscillation between classical and modernist canons that seemed so much the issue in the late 1970s was to evolve rapidly into anxious investigations of a conceivable noncanonical architecture. Polemics between Whites and Grays yielded to a far more direct and devastating confrontation with the rout of modernism—not through the proliferation of its sterile or bastardized versions, or even the revolution in our reading of the avant-garde movements—but from within its indelible richness. For it might be argued that with Wright's Guggenheim Museum and Le Corbusier's Ronchamp the architectural object had already turned resolutely and definitively sui generis. Nonetheless, some of the most provocative speculative projects over the last fifteen years have consisted precisely in demonstrations of this new deregulated condition. Daniel Libeskind's collages, Bernard Tschumi's dismantling, Coop Himmelb(l)au instabilities and contradictions, OMA's darkly witty allegories, and Peter Eisenman's flatbed superimpositions all create the impression of an architecture that wants out and is in despair of finding a viable exit. All of them seem to be raising the elemental question of what an edifice could possibly be now, what kinds of concerns or extrinsic concepts might conceivably legitimate the generation of form.

We make a distinction between architecture and landscape, a distinction that would appear inescapable. This condition of adjacency promotes to this day unqualified assertions that they each participate in irreconcilable orders. To be sure, the architectural treatises have traditionally provided for neighborly negotiations, modulations of the edifice in view of a disparate environment: the perennially recast recognition of site-specificity. But no amount of consecrated green space—not SITE's Forest Building, not Le Corbusier's Plan Voisin or Ville Radieuse, not Wright's Fallingwater or Broadacre City, not the Garden City or Olmsted's parks, not Kew, Versailles, the Palladian villa, or the Renaissance and Baroque gardens, not even Buontalenti's most elaborate version at Pratolino—has ever been given license to invade the tectonic integrity of the edifice. A troglodyte mentality figures as essentially alien to a tradition that posits strict perimeters for its vehicles of high significance. But, between Western metaphysics' considerable investment in positive aboveground building—which finds its most distilled expression in Hegel's *Aesthetics*—and a tradition of architectural treatises that has invariably grounded the validation of the orders in natural laws, there exists a paradoxical neglect of the materia prima that serves as the seat and the resources for all construction. The displacement of the naturalistic paradigm to the human body, like the prestige of the archetype accorded to the primitive hut, reposes on the same quid pro quo, in that each promotes a psychic screening of architecture's fundamental ablation of Earth.

Read Ambasz and note, in his lexicon of predilection, an avowed conception of architecture as a poetics, constitutive of mythmaking and quotidian rituals, ceremonies, and processions, dedicated to what architects like to call the "myth of Arcadia." Taken at face value in a period of fierce demythologizing and demystifcation, these reference points verge on an anachronic preciosity. They have no doubt dampened the interest of more than one historian or critic and led several of Ambasz's more literal-minded commentators to advance the notion that he might be, at once, Wordsworth and Magritte, Ledoux and Jeff Koons, a family-oriented Nietzsche and a New Age Joseph Campbell. Now look at the photographed models, whose carefully composed graphic significance is manifestly at odds with any ultimate realization. The projects are posed in radical isolation, overwhelmed by a preponderance of sky and earth. Only Ettore Sottsass has had the perspicacity to remark that Ambasz's Earth is not at all the picturesque botanical compendium of the pastoral.[5] There is nothing Romantic about this green desert, this turf or tegument stretching to infinity. No historical inflection impinges on the ambient pregnancy released uniquely by the incidence of architectural intervention. Earth, sky—now bring in, as Ambasz's statements tend to do, the personalized rites that inform daily routine, the local deities that reign over place, and consider to what extent these coordinates correspond to those that orient the Greek temple[6] or the bridge[7] in Heidegger's elaboration of "authentic dwelling."

A chasm separates the strict Heideggerian doctrine of the ongoing historical production of meaning from the existential reception that architecture theorists have reserved for it.[8] And the errant path of Heidegger's resonant phraseology passes through the latter's humanistic screen directly into "Emilio's Folly," where the philosopher's better-known philosophemes find their most literal construction. Within the speculative genre of the folly, Ambasz presents a design that unambiguously and unconditionally brings forth the earth, not merely by burying the canonical Mediterranean villa and thereby elevating the surrounding grounds, but, with demonstrative emphasis, by inserting into the villa's courtyard pool an incongruous micromountain, visibly excised from the nearby terrain. To this enigmatic terra incognita he appends a narration or, as he would have it, a "fable,"[9] which foregrounds architecture's specifically discursive feature as it relates, in a first-person voice, a "thrownness" into this preexisting geotectural situation and the consequent progressive personal mapping and appropriation of "what is at hand." No socius legitimates the emergence of significance here, as it must for Heidegger, nor does the architectural object relay attention to the social and historical rootedness of the site. Nonetheless, the signal act of disclosure, animated by care for the successively revealed facets and usages of this location, as well as by a self-reflexive recognition of its temporal dimension, incontestably opens a world. And that openness, figured both topographically and phenomenologically, would appear to infuse Ambasz's poetics with the more compelling semantic force of poiesis: the distinctly human agency involved in the construction of meaningful presence. In the folly's highly synoptic and playful version, however, the combat that Heidegger posits between the world and the earth, that is, between the infinite range of possible human meanings and the inherent constraints of the natural environment (which includes all preexisting worlds) — both largely concealed, each dependent on the other for its historical revelation — becomes the model for an architecture that will take à la lettre the terms *world* and *earth* and afford to their relative concealment and unconcealment the enhanced relief of a rhetorical figure.

The emergence of the earth over and against architecture produces a singularly unfamiliar topography in Ambasz's work, one that scuttles the customary conjunction of solids and voids, and topples the generally recognized hierarchy of primacy and secondariness. Such a deployment of architecture and landscape has been difficult to characterize, because it has been difficult or seemingly impossible to historicize. Only ancient astronomical observatories, Japanese gardens, labyrinths, processionals, and oases offer similarly unclassifiable, hybrid configurations. Far from a territorialization of the site — whether construed as a friendly regionalism or an ominous Heideggerian descent into native soil and the destiny of a people — Ambasz's supraarchitectural conception evokes instead the transtemporal, transcultural enigma of a paleolithic vestige: dislocated, anonymous, without particularities. Like the ruin dating from a time before architecture, which incorporates or recapitulates the natural features of a tract of ground, his structures engage the dynamics of the hidden and the revealed with a categorical shift that overwhelmingly privileges earth surface. On this point Ambasz has been clear. As early as the Casa de Retiro Espiritual, north of Seville, the explicit issue was "to eliminate architecture."[10]

It is important to take the measure of this assault on the seemingly ineradicable gestalt associated with the architectural object, especially in view of the tortuous recent history of the notion of the frame and its implications for the kind of unified and finalized object that architecture implies. If Heidegger had allowed that a constellation of meanings could be harnessed by a certain stability or enframing, say, in the work of art (at least until the paradigms of modern technoscience had rigidified the dynamics of meaning to the point of reification), Derrida later detected the cheating, principally in the overarching designs of philosophical systems, that had been going on all along. What he designated the parergon and was to become the springboard for Peter Eisenman's Carnegie-Mellon Center (1987–1988) represents an accretive frame that results from the remedial plugging of conceptual holes. Such as Derrida describes it in *The Truth in Painting*,[11] the apparently extrinsic parergon is a threatening, critical "additive," a stopgap, a "prosthesis," "half-oeuvre, half-hors d'oeuvre," neither simply outside nor simply inside, detachable but difficult to detach, riveted as it is to what is missing. Countering this ideational assemblage, the text decrees a "general law": make the frame crack. And, indeed, it is in adherence to this law or, more precisely, its image, that Eisenman tackles the problem of designing an edifice that exhibits its own disruptions, disjunctions, and instabilities. To generate the form of the Center, he conceives of the frame as a linear cube, projects it in n-dimensions, cuts it axially, and rotates the segments, seeking the cube's complex, virtual relations.[12] But Eisenman's choice of the cube as a point of departure tells the whole story and returns us to square one. The end product of this process, for all of its formal breaks and lattice extensions, constitutes a necessarily static, constructable unit, which may illustrate its generative principle, but amounts in toto to nothing more than a reframing. And reframing, together with the dream of no frame at all, are the two contradictory gestures that Derrida has labored so hard to deconstruct.

The receptacle for habitation shows a remarkable resistance to contemporary theories of process and dynamics. No obsolete metaphysical imperative, the sheer material necessity for closure and immobility has been problematized in our time far beyond anything Bruno Zevi might have dreamed of in extolling the traditionally neat, invaginate relations between the shell and core of a solitary edifice. To the dilemma of introducing a kinematics into architecture's statics, Ambasz responds with an unequivocal negation of tectonic symbolization, in its place consigning the edifice to literal interment and leaving "process" to the surrounding soil. A thorny question arises, then, as to just what kind of frame landscape might provide, and nothing is more ambiguous. In Ambasz's projects, a strict heteromorphism exists between the habitation and its landscape boundaries; their respective contours remain distinct and arbitrary. The clearing that landscape affords the site predominates, surrounding the habitation, submerging and sometimes invading it with the material substance of an improper territory. As a frame, the terrain is bivalent: while it surely serves to harness the site, at the same time, it opens it up, allowing it to expand and dissolve into the wider surroundings. When the philosophical text overflows and cracks apart, Derrida was to remark, it is condemned to find only other texts.[13] When the architectural object overflows and cracks apart—the accumulated history of architectural models notwithstanding—ultimately it joins the natural continuum, of which it is merely one type of articulation.

Ambasz's grafting of landscape onto architecture retains an outlandish aspect only so long as architecture maintains its status, according to the habitually parochial view, as an autonomous enterprise cut off from simultaneous permutations in the visual arts. In New York in the late 1960s and into the '70s, where and when Ambasz's ideas were taking shape, boundaries were becoming exceptionally permeable. Painting had already come off the wall in a full recognition of its three-dimensionality, and sculpture was taking a decidedly unclassifiable turn. As Rosalind Krauss was to sum it all up at the end of the decade, the hitherto freestanding object had surprisingly invaded formerly unoccupied regions delimited by the normally extrinsic architecture and landscape, positively or negatively defined: landscape and architecture, not-landscape and not-architecture, landscape and not-landscape, and architecture and not-architecture.[14] Crossbreeding produced hybrids: site-constructions; earth mounds and excavations; marked sites, inserting man-made natural structures into the natural environment; and axiomatic structures, seeking to simulate aspects of architectural experience. In the mapping of this logical expansion, Krauss recognized the "lifting of an ideological prohibition to think the complex," which represented an unequivocal "historical rupture." What is astonishing, in retrospect, is not only that she should have bolted her argument to evidence that "other cultures had been able to think the complex," specifically identifying examples of both landscape and architecture (mazes and labyrinths, Japanese gardens and ancient processionals), but that the reciprocal implications of this "historical rupture" for architecture itself was to remain obscure. Seen today in the context of Ambasz's work, Michael Heizer's *Two-Stage Liner Buried in Earth and Snow* (1967) and *Compression Line* (1968) or Robert Smithson's *Partially Buried Wood Shed* (1970) look like rough preliminary drafts for an integral architectural program. For the underlying urge, as Smithson emphasized in his writings, was to arrive at something "sprawling and embedded in landscape, rather than putting an object on the landscape."[15]

Minimalist art has occasionally been evoked to characterize Ambasz's use of simple geometric forms. But the minimalist mindset, pushing ineluctably toward the earthwork mutation, had a much more far-reaching and subversive objective: a total rejection of the work's inner logic. The black boxes, L beams, cages, and floor tiles established an art object that was nonrelational, opaque, and external. All that mattered was the materiality of the object, its mass, bulk, density, and volume. The meaningful core, the composition, the interior had been leveled to an inexpressive, enigmatic surface that, in its tension between self-evidence and impenetrability, converts the viewer's experience of a shared space into an inhabitual confrontation with what Krauss so rightly designated, with regard to LeWitt and his peers, as aporia.[16] Consequently, the kind of inscrutable found object that the minimalists were trying to achieve finds its correlative in Ambasz's projects, in point of fact, not at all in his use of simple geometric shapes, but rather in his appropriation of landscape. Once you have buried the box (as LeWitt did, quite literally) or consigned the receptacle for habitation to some arbitrary location, lurking in the bowels of Earth in an indeterminate way, the issue of the architectural object's inner logic is dead. The work turns centrifugal: its familiar natural surface, defamiliarized by an indecipherable architectural organization, becomes a marked point of fascinating estrangement. As architecture, landscape, in its striking inevidence, stares back at the observer with dumbfounding undecidability.

Ambasz's purism, a purism that often reflects a Scarpa-like refinement in the articulation of detail, studs a much larger design. When visible, the strictly architectural fragments—an undulating wall, the intersection of two planes, a beveled sphere, receding tiers, a curvilinear stairway, a trellis or lattice—obey a principle of extreme reduction, even as they establish a rudimentary signaletics of the site. What is being carried out is something on the order of a demonstration of the least number of units required to mean architecture, not unlike those employed by the Mexican architect Luis Barragán, whose work Ambasz was to unearth and exhibit when he was curator of design at the Museum of Modern Art in New York. Master of the isolated architectural element—the gate, the reflecting pool, most notably, the self-supporting wall—Barragán proposed a form of architecture that verges on both sculpture and scenic design. His use of brilliant color and the backdrop of a raw and antithetical Mexican landscape in which his elements acquire enhanced relief, however, serve only to augment the assertive force of his structures as autonomous modernist works of art. Barragán's work, in its way, is all purism, whereas Ambasz's version takes on its specific resonance as a counterpoint to landscape's predominating effect of contamination. If an "impure-purity," as Smithson maintained, was to become the hallmark of visual art in the late 1960s and early '70s, clearly in the wake of a renewed Duchampian invitation to spoliate a presumably pristine modernism, in architecture this tendency was to take the form of recourse to vernaculars and revivals. Smithson wrote of his generation's enchantment with "the mood of bad architecture"[17] and the transtemporal richness of 1930s Ultramoderne eclecticism[18] some time before Robert Venturi, Charles Jencks, or Paolo Portoghesi turned their attention to those wellsprings. And it would be left primarily to the visual artists of this generation to extrapolate the implications of earth-surface and seize upon the promenade architecturale as their own veritable medium.

A house conspicuous by its absence troubles the imagination. In Ambasz's most prototypal residential designs, the House for Leo Castelli, Manoir d'Angoussart, and La Casa de Retiro Espiritual, something habitual in the habitation has been wrested from the site, and what is left belongs to a rhetorical topos hotly debated in France during the period of their inception: the rest. The rest is a throwaway in underground works like the Philip Johnson Gallery or Norman Foster's bunker house. Yet the rest was traditionally a regulator in the definition of the suburban or country villa—a regulator later perceived as a source of suspicious deflection. As Wölfflin remarks laconically in *Renaissance and Baroque,* the crossing of art and countryside seems to take the art out of art.[19] And, in much the same vein of urban reticence, Tafuri continues: "In the landscape context, from an absolute object architecture becomes a relative value."[20] But, indeed, it is just this relativity, implicit in the conception of the suburban dwelling from Martial and Terence to Alberti, that contributes to a principle of intervention into landscape requiring development, not in height, but at ground level and out into a larger promenade architecturale. The intriguing issue that Ambasz raises, then, pertains to the purview of that principle, that is, the extent to which the

House for Leo Castelli, East Hampton, New York, USA, 1980

architectural work might be conceived in its entirety as a promenade, in and around and even over the house. Bird's-eye views of his designs—a vantage for which he appears to have particular affection—suggest precisely this kind of unified, all-over surface, guided by sculpted roads and extruding markers. "The landscape," much the way Smithson imagines it in his proposal for an "Aerial Art," "begins to look more like a three-dimensional map than a rustic garden."[21]

Ambasz's early professional association in Buenos Aires with Amancio Williams alone would have sufficed to acquaint him with an acute Wright-inspired concern with landscape. Williams's Casa del Puente (1943–1946), for example, engulfed by an overgrowth of lush Mar del Plata vegetation, assumes the posture of an enclosed bridge that straddles the river, bolstered by a frank curvilinear support. But, quite apart from his reputed accessory obsession with horizon lines, Williams was to oversee the construction of a design that contains in nuce an assortment of architectural impulses recognizable in Ambasz's disparate but often repeated gestures: Le Corbusier's sole work in South America, La Casa Curutchet (1949) in La Plata. At street level, this suburban villa distinguishes itself notably by a detached, prismatic portal. The four stories of the structure, accessible via interior ramps, however, closely encase a foreign body: a fully mature tree, rooted at ground level and sprouting upward past openings in the facade and beyond the roof, like a mammoth, domestic potted plant. That is to say, this arborescent hostage occupies an open core around which the promenade architecturale leads with its unraveling logic up to a baldachin-covered approach to a terrace-garden, overlooking a green panorama. Within the continuity of

Manoir d'Angoussart, Charleroi, Belgium, 1980

tree, garden, and panorama, the axis of the villa's alternative facade shifts the garden door into the position of a second principal entrance, allowing the promenade to function both top down and bottom up. Here, on high, it becomes entirely possible to imagine the application of a strict rule of architectural nonnecessity, much as Ambasz has done, which would lift the terrain to the terrace level and sink the villa out of sight.

If the impulse, then, is to achieve extension without interruption, orientation within the architectural work will depend upon a set of signals that consist of architectural fragments. Like a latter-day Piranesi detailing sections wrenched from the ruins of modernism expressly for an excavationlike installation, Ambasz elaborates a series of detached, isolated, incomplete forms that point indexically to the physical presence of an unapparent buried structure. At the same time, the resonance of each extrusion and each void fans out, effacing any notion of a perimeter or end to the work, thereby foregrounding the underlying natural continuum. Gordon Matta-Clark's excisions and microdemolitions in existing architectural structures, which encourage reciprocal relations between inside and outside, between spatially and temporally disparate cultural fabrics, were typical of a postminimalist concern with the capacity of the work to absorb and to be absorbed by a preexisting physical environment. Ambasz's interpretation of this concern for architecture can best be seen in the constructed Private Estate in Montana, gobbled up as it is by the real brute terrain, a steep slope from a dense pine forest down to a lake. Pierced with oversized windows, the facade stands alone in its elegant, antithetical concavity, more void than solid, an annunciatory or exclamatory marker in the wilderness. Excavation miraculously deromanticizes Arcadia. Eruptions from below, those chthonic forces thought to rise and haunt, were devaluated and evacuated by positivistic scrutiny. In Earth's elevations and depressions, Darwin and De Beaumont found a different sort of "legend" by which to gauge the hidden, sinking strata from above.[22] We read Earth through its geomorphic plan, the way we read the body through its skin.

Ambasz's hidden corpus reawakens what might otherwise be taken as the stale analogy or homology between the edifice and the human form. Philosophically charged, yet curiously unsubstantiated in its Greco-Roman and Renaissance versions, the model and even the later Modular seem to have withered away. Now architectural truncations of all descriptions suggest contemporary resurgences, with the Lacanian morselated body as a facile guarantee. What lives and breathes in landscape is the green tegument, the expanse of dermis, the long-neglected, underestimated, undervalued organ called the skin. The modernist glass sheath seems a thin and brittle metaphor in comparison with this geological metaphor for the cutaneous, unpeeled and elastic, neither merely outside nor merely inside, ectomorphically one with the brain, nerves, and sensory organs—quite literally, as Ambasz employs it, an immunological layer, a thermo, hydro-regulator, its orifices communicants with the interior milieu. Look at the villas, the Schlumberger Research Laboratories, or the Lucile Halsell Conservatory and remember Walter Benjamin's detection of a shift in the artist's identity—picked up by Tafuri as so pertinent to architecture—from magician to surgeon.[23] However much Ambasz has been associated with the abracadabra of his mists and vapors, clouds and odoriferous shrubs, it is hard not to see in these schemata operations of the scalpel's deft effect, the clean incision. Engraved here is no anthropomorphism, but rather passages traced for bodily experience, which transits with phenomenological drive through a preeminently material world.

Neat carvings in the earth, linear inscriptions, divulge the composition of Ambasz's berms. What is a berm if not the edge, the ledge, the limit of some larger surface, taken as primary? From this primary solid, the habitation becomes negative void. Berms and grottoes, caves and bunkers, subterranean temples, catacombs, mines, and basement playrooms, vaults of the modern urban infrastructure (to which Ambasz would write an early encomium), not to speak of the effervescent artistic and political hollows of underground activity,

Schlumberger Research Laboratories, Austin, Texas, USA, 1982

attest to an atavistic sense of this space's cultural availability. But the issue of "The Burrow," in Kafka's penetrating analysis,[24] swings gravely between a second skin for the peaceful fantasies of intrauterine repose and a labor-intensive crypt fit only for apprehensiveness and terminal anxiety. Being below ground level invites this uneasy acknowledgment of womb and burial. Cut any of Ambasz's structures axonometrically and find the paradigm: the burial mound, the tumulus. Slice a tumulus in a similar fashion, however, and find a not-uninviting abode, quiet and impregnable. This bivalence, of course, informs all architecture. Any hole in the wall, as Bachelard observed, can become an insular sanctuary, just as architectural aspiration, to believe Boullée and Adolf Loos, finds its apotheosis in the funerary monument. The quintessential cenotaph, as formulated in Boullée's *architecture ensevelie,* reduces architecture to its bare essentials, its skeleton, a naked wall without a plinth. Far from the hermetically sealed enclosure, however, Ambasz's berms surprise by their openness. Every edge and every ledge reiterates the geological phenomenon of an open-air depression in the earth. At bottom, the plan for the houses reinstates the model of the Mediterranean villa, whose rows of rooms open onto an inner court. Sources of ingress and egress, light and ventilation, repeatedly pierce the buried structures with unexpected stairs and skylights, or the gaping glass embrasure of a doubly exposed entrance-reception room. Claustrophobia and the fear of asphyxiation cede in the villas—which force the outside in and inside out, establishing a delicate balance between protection and exposure—to a much more complex phenomenological experience of the architectural opening toward the agora, wide-open space, green prospects sometimes within and always just beyond the roof. But the berm's outward thrust involves as well a deviation from the habitation's ordinary rootedness. Given the disjunction between the habitation's form and the landscape configuration, the superficially buried cubicula could multiply under the earth in any direction, its segments in no way regulated by the hierarchy implicit in the central trunk of a conventional edifice. With no beginning and no end, anchored in the earth by a minimal foundation, so like a foundation in itself, this root canal inevitably sprouts upward—a detached facade, a mirador, a solar energy unit, a ventilation duct, a rooftop rim—like an orchid, an iris, a dahlia. Deleuze and Guattari's figure of the rhizome, the loosely planted, horizontal, nomadic root effects a shift in logic from the Western forest to the Oriental steppe or garden, desert or oasis, in which the cultivation of the tuber fosters the development of more tenuous and unpredictable relations. "In the last analysis," they were to say, "it is always the grass that has the final word."[25]

    What is commonly referred to as the fusion between architecture and landscape in the garden provides a necessary but insufficient topos for Ambasz's procedure. Ultimately, the question is not the extent to which architecture has been given free rein to design a highly wrought, quasi-autonomous, artificial natural environment, but the extent to which this natural environment can be deployed as the gestalt itself of the primary architectural object. For this, there are no clear antecedents other than the idiosyncratic cases of Taut's speculative Alpine Architecture or César Manrique's constructions in the Canary Islands. The baroque imitation of the organic in stone suggests a similar impulse; yet, in its very mimetic principle, it turns antithetical. Only Buontalenti's adventure at Pratolino fully embodies the effort, not merely to implant tectonic elements in the landscape or to organize the landscape in a tectonic manner, but to create an artificial universe of forms, sounds, and visual effects, purposefully variegated and irregular in its overall design, out of brute matter. Where the rationalism of the Medici court that carried the desire for urban evasion toward an extensive investigation of available botanical, hydraulic, and mechanical techniques for expanding the definition of landscape brings the enterprise to its full achievement—and abrupt halt—is just beyond the southern entrance to the villa. Situated in a lower-level terrace, the multichambered grotto, which merited a detailed description of its movable decors, automata, and cascade music in Montaigne's *Journal du voyage,* in no way impacts on the especially austere villa that stolidly contains it. Buontalenti's Pratolino was nonetheless to remain an object lesson in the garden's formal potential, just as the legacy of the Renaissance and Baroque gardens persists in a summary vocabulary, each term of which figures conspicuously in Ambasz's work:

Axes. Ambasz's incisions, his edges and ledges, follow a remarkably consistent pattern: almost invariably straight lines that meet at right angles, counterpointed by an undulating curvilinear stroke. At most, the undulation expands into a hemisphere or circle. This highly reduced, abstract script, as in the plan for the Schlumberger Research Laboratories—a site plan exceptionally loose in organization, potentially extensible or modulable, as are the detachable,

Mercedes-Benz Showroom, New Jersey, USA, 1985

movable individual units buried below — effects a singular topographical inversion: pedestrian paths on the expanse of grass-covered terrain are left unmarked, whereas the tectonic elements instead assume the forms typical of the garden axes. All of the villas show a similar configuration of pedestrian freedom and axes solidified into architectural components. Emblematic of this pattern, the Mercedes-Benz Showroom, most appositely, consists in nothing more than the most basic consolidation of a crossroad or axial intersection.

Pergolas. In a similar reduction seen in Ambasz's project for the Cooperative of Mexican-American Grape Growers, the pergola, together with its attendant fruits, makes a claim to architectural self-sufficiency. As a more conventionally integrated structure in the plan for Frankfurt's Eschenheim Tower, the arbor provides an openwork enclosure for urban connections in a way that recalls Hector Horeau's projects for covered avenues in Paris. The point is not obscure: the trellis/lattice/grid plays on a homology between armature and vine and, at the same time, maps a space, shuttling it back and forth between the determinate and its implicit infinite extension. In this way, Manoir d'Angoussart's ghostly trellis facade announces entry into a locus solus that elicits the sense of a phantom habitation floating over its material double, one that has slipped out of view into invisible subterranean reaches. An uncomplicated combinatorial process makes of the trellis and the garden axes the constitutive elements of the Nishiyachiyo New Town Center, emphasizing their sufficiency even in the radically isolated, monumental proportions of a town center in an as-yet-undeveloped region. The grid, by turns angular and curvilinear, void on one side, solid on the other, plots an implicity open perimeter, studded serially with the garden's proverbial potted flowering tree. Railway axes orient the design toward two thirty-five-story office buildings in the guise of a traditional Oriental Torii Gate.

Pools: The sunken central courts of the Casa de Retiro Espiritual, the House for Leo Castelli, or the Phoenix Museum of History in no way establish a bottom. A reflecting pool, often amorphously shaped,

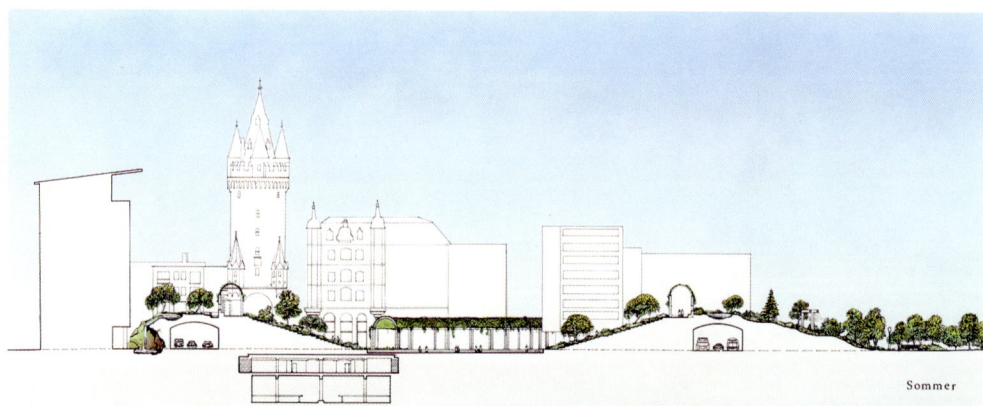

Eschenheim Tower, Frankfurt, Germany, 1985

deflects from the strict delineation of the dwelling proper, directs the promenade, and intimates unspecified further depths, both downward and upward, in its reflection of sky and contingent resurgences of the surrounding landscape. Thus, the swimming pool technology that assures the berm's feasibility also assures a rhetorical function. At Casa Canales a triangular rooftop pool camouflages the house below and and emphasizes both contiguity and continuity with its containment of a patch of mirrored firmament. At the Schlumberger Research Center, the body of water meanders, yet not without further Italianate accessories: the bridge and isolotto. The sinking remnant of land mass, in its imaginative, artificial version, constitutes a key chapter in Ambasz's "scientific autobiography." His inaugural high-tech workstation barges at the Center for Applied Computer Research, perhaps with a slight bow to Le Corbusier's floating Asile, have been relaunched on several occasions and for less self-exegetical purposes than the allegorical mound in "Emilio's Folly": the floating galleries for the New Orleans Museum of Art, the enlargement of preexisting bodies of water at Paseo del Lago or the Shikoku Marine Resort Community provide an elemental, liquid trope available for repetition in an accessory horizontal rhizome-like motif within the wider context of site-specificity and land development. The progressive elaboration of this liquidity brings the isolated cascade at the Grand Rapids Art Museum, as well as the seemingly gratuitous cascade and pool at Houston Center Plaza, into relief as historically grounded garden motifs.

Grottoes. The significance of an axonometric cut in an edifice buried under a layer of topsoil was not unknown to Enlightenment architects. What you have, at rock bottom, is a grotto. No amount of civilization can dislodge this void as a mythic locus of habitation. Nor can the most ponderous work of architecture avoid the interior void it must, of necessity, define. Ambasz's excavations represent mock grottoes that stress an empty core within a first principle of terrestrial or subterranean substantiality. In this way, the burrowing of the berm offers an escape route from architecture's historical veneer of solidity. When called upon to intervene inside a preexisting empty core, however, Ambasz once again inverts the formula. His proposal for the reuse of Union Station presents a scene that at first glance might be taken for a cataclysmal eruption of landscape from out of the structure's bowels into its poised gaping Great Hall. In its assertion of a contaminating force, this invasion summons up the image of Brongniart's Project for a Mountain in Saint Andre Cathedral. The breach of decorum sought by French Revolutionary secularization finds something of an aesthetic equivalent in this geological insurrection against architectural language tout court—a stand reinforced at Union Station by a design that appears to urge the adjacent Liberty Park to rush down a steep drop, like a cataract, and flood these academic precincts, seeping deep into the building's lower levels, nearly rejoining the earth below. As solutions to outdoor urban reconversion, Houston Center Plaza and Plaza Mayor both insert public space within a subterranean cavity that converges with Ambasz's villa prototype. At Plaza Mayor, however, the requisite ground-level green carpet has been tufted by a tree-lined parterre.

Terraces & Stairs. Ground level is relative, easily displaced. Ambasz's descents might appear timid when compared with Le Vau's exceptionally elastic conception of the light and air available in great depth: a 330-tiered orangerie, for example, set beneath the terrace of a seventeenth-century chateau. From the moment the garden terrace dismissed the lay of the land and, with agronomical authority, reproduced its layers in ascending or descending tiers, the plan in extension was turned on its side. The garden became a vertical enterprise: terraces, tiers, the place a gradin (theater, amphitheater, circus), the step, the stair—the ziggurat. With Ambasz's emergence from below stairs into freestanding structures, the plateaux have simply multiplied and mounted. What is the Fukuoka Prefectural International Hall if not fifteen one-story receding terraces, that is to say, a monumental stairway? It would be reasonable to see the cascade flowing down to a theater below as a playful wink at the Villa Aldobrandini and accept the extension of the preexisting park, up and over the building out to the sea and sky, as merely one more example of Ambasz's obsession with landscape continuity. But the dramatic accentuation of an essentially heavy, truncated, stepped-back form forces an association with the turning point in the deployment of the garden-stair: Bramante's Belvedere at the Vatican. Like Bramante, Ambasz employs terraces and stairs—theater below, belvedere above—to articulate the gap between two preexisting architectural blocks. His version, however, is detached, literal, repetitive, in a way that conforms to a contemporary minimalist or rhizomatic logic: segmentation, stratification, one thing after another. "A plateau,"

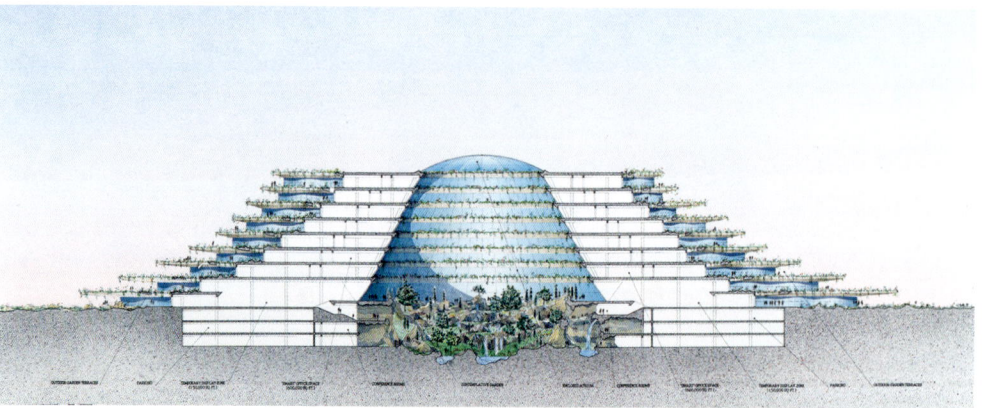

Worldbridge Trade and Investment Center, Baltimore, Maryland, USA, 1989

as Deleuze and Guattari insisted, "is always in the middle, neither beginning nor end. A rhizome is made of plateaux."[26]

Under the foliage, a plateau: Deleuze and Guattari's "Rhizome" was conceived as a position paper for an alternative philosophical discourse, no longer systematic or unified, but infinitely extensible by way of tangents: *mille plateaux*. Derrida surely figures as the signal reference point. His repeated, fragmented, differential reentries into the seminal texts of Western metaphysics is revealingly summed up in a single phrase from *The Truth in Painting:* "I do not know what is essential or what is accessory."[27] The point is graphically reinforced by illustrations of Fantuzzi's ornate, self-proliferative frames that become the work itself around an empty center. Taking an impulse discernable in Lequeu's Sepulture near Voorhout to an extreme point of integration, the site plan for Ambasz's Worldbridge Trade and Investment Center shows an expansive indeterminacy that dilates any accepted architectural notion of a shell, spreads it out in amorphous lateral and centrifugal botany-encrusted layers from a hollow core to the virtually undifferentiable landscape. Only the free-form surrounding clearing detaches the structure. With its echoes of Boullée's cenotaphic enceinte, the truncated conical core refutes its centrality as the hold of habitation, which has been shifted to a peripheral interior. What emerges at Worldbridge is not merely multiplied earth surfaces. Their irregular outer contours, which could be taken for the terraces of the tilled field, coalesce into an unmistakable protozoan form. With this instance of zoomorphism, not to speak of the gigantic entrance tear at Fukuoka, as if the building were made of natural tissue, a landscape premise predicated on the literal infusion of organic matter into architecture has been reformulated to admit symbolic reiteration.

Greenhouses. Organicism is both the bête noire and the holy ghost of late-twentieth-century thinking. The naturalistic metaphor of the rhizome surprises. Yet the same Smithson who had denounced Wright's Guggenheim Museum as an "inverted intestine," a "concrete stomach,"[28] filled with a suspicious and obsolete anthropomorphism would later stare into his own earthwork, *The Spiral Jetty,* and observe: "Following the spiral steps we return to our origins, back to some pulpy protoplasm. [...] I was slipping out of myself again, dissolving into a unicellular beginning, trying to locate the nucleus at the end of the spiral."[29] The shift in focus in the visual arts from organicism to physical science goes awry, attesting in the end only to unsettled scores with natural phenomena. For Earth is a powerful vehicle: a tough membrane, most easily likened to a single, living cell. Ambasz was to receive his first green light for construction with the Lucile Halsell Conservatory. In San Antonio, the earth's insulating capacity made climatic sense, and Ambasz's high-tech peaks of glass and steel not only rise as sleek variant indices of sublevel structure but concretize his unequivocal relation to an auxiliary architectural tradition: the greenhouse. All of the unsettling ramifications of Ambasz's extrapolations of organic matter were to have found initial acceptance, ironically enough, within the context of vegetation in captivity.

The greenhouse microclimate takes on a particularly enigmatic dimension at the Nichii Obihiro Department Store: a two-and-a-half-acre building draped in greenery and encased in a confoundingly irregular, multifaceted bell jar. This green promontory or mountain-mall on the Siberia-like Hokkaido Island flirts with mimesis, even though its organicism and inorganicism reference no known geological formation—that is, apart from the entry to the cyberspace Free Zone in William Gibson's *Neuromancer*[30]: the interior walkway through a deep canyon; the subtle angles of the boutiques and enclosures that dissimulate its walls; the light from above that filters through masses of green vegetation cascading down from numberless

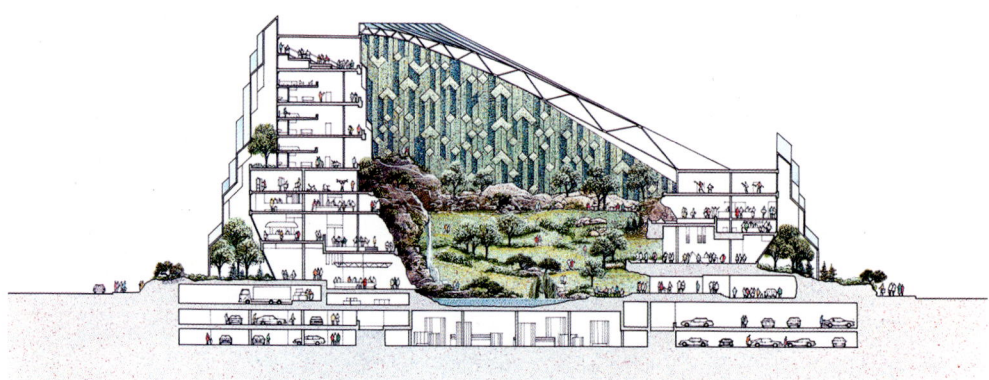

Nichii Obihiro Department Store, Hokkaido, Japan, 1987

terraces and balconies. Ambasz's state-of-the-art transplant, however, reverses this anomalous green valley core in the overarching exterior configuration of an equally anomalous green mountain peak, which exasperates (to use Tafuri's term) the greenhouse genre by forcing a Taut-like Alpine flashing of glass in the crystalline environs to contain a visible chunk of disparate and thoroughly artificial geography. Smithson warned of the mind-bending and unnatural associations produced by "quality gardens."[31] With this extravagant example of hothouse domestication, Ambasz's interpretation of Louis Kahn's wrappings veers toward the purely rhetorical statement, even as the structure's hermetic closure instills a disturbing sense of our estrangement from the material substance kept in confinement inside.

"Nature" has not yet been evoked, not once, and for good reason. Here it emerges, harnessed by quotation marks, a citation of an antiquated common place. Facile references to this historically eminent topos, a frequent, mindless reflex, have gone far to obscure the thrust and timeliness of Ambasz's work. Is landscape "Nature"? An abyss opens, in this spot, between sheer physical matter and the universal history of human concepts and techniques. Were landscape in its symbolizing power still an adequate vehicle for some natural totality, it would be nonetheless legitimate to ask: what "Nature"? To confine the question only to the West: Pre-Socratic first principles? The Greek physis? The Roman natural? The alternatively secularizing, moralizing, or redemptive Renaissance "Book of Nature"? The object of Copernican or Newtonian speculation? A Leibnizian monad? Kant's sublime unknowable? Enlightenment natural laws? Hegel's suicide victim? Romantic poetry's visionary resource? Positivism's captive? Bergson's élan vital? All of this and more haunt the generic noun, filling it with an unsuspected heterogeneous surfeit of meaning to the point of utter opacity. But the opacity, the illegitimacy, the pure emptiness of landscape, as a statement in itself, is exactly what defines Ambasz's architecture as so pointedly contemporary. At a time when architectural demonstrations of depletion of meaning, rebuttals, and refusals of signficance prevail, landscape shows itself to possess a unique capacity to store its unparalleled illustrious connotations, flirt with them, subvert them, and to persist, even in its stark semantic nakedness, with untarnished autonomy.

To take Ambasz's architecture as resolutely nonurban—a throwback to Wright's disgust with the metropolis and his search for agrarian alternatives—would be to ignore a dramatic alteration in cultural options and sources of value. Surely the parallels are real: the "fusion" of architecture with landscape, the influence of the artifical Italian garden, the reliance on technology to make the earth a viable habitat, and even the penchant for what Mumford would call "solo performances." Still, in Ambasz's case, the urban/nonurban distinction

An earlier version of this essay was published in 1994.

1 Manfredo Tafuri, *La sfera e il labirinto: Avanguardie e architettura da Piranesi agli anni '70* (Turin: Einaudi, 1980), 365.
2 Manfredo Tafuri, *Théories et histoire de l'architecture*, trans. SADG (Paris: SADG, 1976), 155. An English translation appeared under the title *Theories and History of Architecture*, trans. Giorgio Verrecchia (New York: Harper & Row, 1980).
3 Tafuri, *La sfera e il labirinto*, 365 (see note 1).
4 Tafuri, *Théories et histoire de l'architecture*, 160 (see note 2).
5 Ettore Sottsass, "Ettore Sottsass, Milan, Italy," in Emilio Ambasz et al., *Emilio Ambasz: The Poetics of the Pragmatic* (New York: Rizzoli, 1988), 10.
6 Martin Heidegger, "L'Origine de l'œuvre d'art," in *Chemins qui mènent nulle part*, trans. W. Brokmeier (Paris: Gallimard, 1962), 32–35.
7 Martin Heidegger, "Batir Habiter Penser," in *Essais et conférences*, trans. Andre Preau (Paris: Tel Gallimard, 1980), 180–85.
8 The reference is of course to Christian Norbert-Schulz, Alberto Perez-Gomez, and, most important, Kenneth Frampton. See Kenneth Frampton, "On Reading Heidegger," in *Theorizing a New Agenda for Architecture: An Anthology of Architecture Theory, 1965–1995* (New York: Princeton Architectural Press, 1996), 440–46; originally printed in 1974.
9 Emilio Ambasz, "Emilio's Folly: Man Is an Island," in *Emilio Ambasz: The Poetics of the Pragmatic*, 162 (see note 5).
10 Emilio Ambasz, "I Ask Myself," in *Emilio Ambasz: The Poetics of the Pragmatic*, 28 (see note 5).
11 Jacques Derrida, "Parergon," in *La Verité en peinture* (Paris: Champs/Flammarion, 1978), 19–168.
12 A parallel might be drawn between Eisenman's examination of the cube and Sol LeWitt's cubic structures of the 1960s. In point of fact, however, the German artist Manfred Mohr had by the mid-1980s established a body of work, initiated in the early 1970s, based specifically on computer projections of the n-dimensional cube in rotation. See my "Linebreeder," in *Algorithmische Arbeiten* (Bottrop: Josef Albers Museum, 1998).
13 Jacques Derrida, "Tympan," in *Marges de la philosophie* (Paris: Minuit, 1972), xx.
14 Rosalind Krauss, "Sculpture in the Expanded Field," in *The Anti-Aesthetic*, ed. Hal Foster (Port Townsend: Bay Press, 1983), 31–42; originally printed in *October* 8 (spring 1979).
15 Robert Smithson, "A Sedimentation of the Mind: Earth Projects," in *The Writings of Robert Smithson*, ed. Nancy Holt (New York: New York University Press, 1979), 89.

seems null and void. His production involves a prototype developed, not in the country, but in total isolation, a pure laboratory product. What it rejects is an evolution of architecture within a typically urban setting, myopic vis-à-vis architecture itself. To this extent, Ambasz's reading of Arcadia, a varied literary genre of urban evasion, sometimes halcyon, often elegiac, turned "myth" only for architecture, would seem much closer to what Smithson was to label "land reclamation." For if the artistic recycling of a spoliated or dilapidated landscape might best be achieved by earthworks, within the context of architecture's habitual ablation of the preexisting site, construction might well be conceived as a preservation or a resurgence or, at the very least, a reemphasis of natural terrain. What has often been construed as a flagrantly ahistorical architecture, oblivious to the site's temporal strata or the architectural language of contiguous elements, responds, in reality, to another equally forceful historical imperative: to move outside just those perennial coordinates.

La Venta is Ambasz's most synoptic project. With an extreme freedom of design that admits the amorphous, the truncated, and the aleatory, the seven buildings instate the same carefully cut, raised-earth-surface, terraced principle. Only ramps to the roof have been added for motorized access and attenuation of any sense of scale. Engulfed in a dense, dead pine forest, these tiers planted with saplings set in motion an odd dialectics of dead and alive. Our capacity to discriminate between the preternatural and the natural, the cultivated and the brute, wavers. Smithson experienced a similar perceptual upheaval when faced with the interpenetration of the dead Great Salt Lake with his *Spiral Jetty:* "My dialectics of site and nonsite

Complejo de Oficinas, La Venta, Mexico, 1993

Complejo de Oficinas, La Venta, Mexico, 1993

whirled into an indeterminate state, where solid and liquid lost themselves in each other. It was as if the mainland oscillated with waves and pulsations, and the lake remained stock-still." At La Venta the animation of the inanimate, however, is no effect of the optical imagination. The dead pine forest, itself man-made, had been asphyxiated by its density. To redress that human miscalculation, Ambasz converts architecture into the cultivated grounds for its regeneration. In consequence, La Venta might be read as "Nature's" cenotaph. The "mythmaking" that he has always seen as architecture's vocation arises, then, in the mythic moment of a historical threshold: the death of "Nature" and the emergence of its complex materiality. Or, perhaps, the death of "Nature" and the advent of the hybrid form: an architecture of the graft, a mutant in the age of genetic interventions, in which human calculation exerts an unimpeded projective power, and the parterre is everywhere.

16 Rosalind Krauss, "LeWitt in Progress," *October* 6 (fall 1978): 60.
17 Robert Smithson, "Entropy and the New Monuments," in *The Writings of Robert Smithson,* 11 (see note 15).
18 Robert Smithson, "Ultramoderne," in *The Writings of Robert Smithson,* 41 (see note 15).
19 Heinrich Wölfflin, *Renaissance et Baroque,* trans. Guy Ballange (Paris: Livre de Poche, 1967), 305.
20 Tafuri, *Théories et histoire de l'architecture,* 114 (see note 2).
21 Robert Smithson, "Aerial Art," in *The Writings of Robert Smithson,* 92 (see note 15).
22 Francois Dagognet, *Une Épistémologie de l'espace concret: Neo-géographie* (Paris: Vrin, 1977). See especially Chapter 2: "Reliefs et paysages, pour une épistémologie de la géomorphologie."
23 Walter Benjamin, "L'œuvre d'art à l'époque de sa reproductibilité technique," in *Poésie et Révolution* (Paris: Denoël, 1971), 196.
24 Franz Kafka, "The Burrow," in *Franz Kafka: The Complete Stories,* trans. Willa and Edwin Muir (New York: Schocken, 1971), 325–59.
25 Gilles Deleuze and Felix Guattari, "Rhizomes," in *Mille Plateaux* (Paris: Minuit, 1980), 28.
26 Ibid., 35.
27 Derrida, "Parergon," 73 (see note 11).
28 Robert Smithson, "Quasi-Infinities and the Waning of Space," in *The Writings of Robert Smithson,* 33 (see note 15).
29 Robert Smithson, "The Spiral Jetty," in *The Writings of Robert Smithson,* 113 (see note 15).
30 William Gibson, *Neuromancer* (New York: Putnam, 1984), part 3, 10.
31 Smithson, "The Spiral Jetty," 91 (see note 15).

I strive for an urban future where you can open your door and walk out directly into a garden, regardless of how high your apartment may be. I submit it that my building in Fukuoka is one example of how we can, within a high-density city, reconcile our need for building shelters with our emotional requirement for greenery. EA

Fukuoka
Prefectural International Hall

Fukuoka, Japan, 1990

The city of Fukuoka in Japan desperately needed a new government office building, yet the only available site was a large park in the center of town. When the news emerged that this potential new structure would be built on the city's last remaining green area, the citizens of Fukuoka erupted in protest. Ambasz was awarded this commission for successfully achieving a reconciliation between the two opposing desires: he doubled the size of the park while also providing the city with a powerful symbolic structure at its center. His design utilizes a series of garden terraces stepping up the facade of the building, thereby giving back to Fukuoka's citizens virtually all of the land that the building would subtract. This met with immediate community approval, thus avoiding any construction delays due to protests.

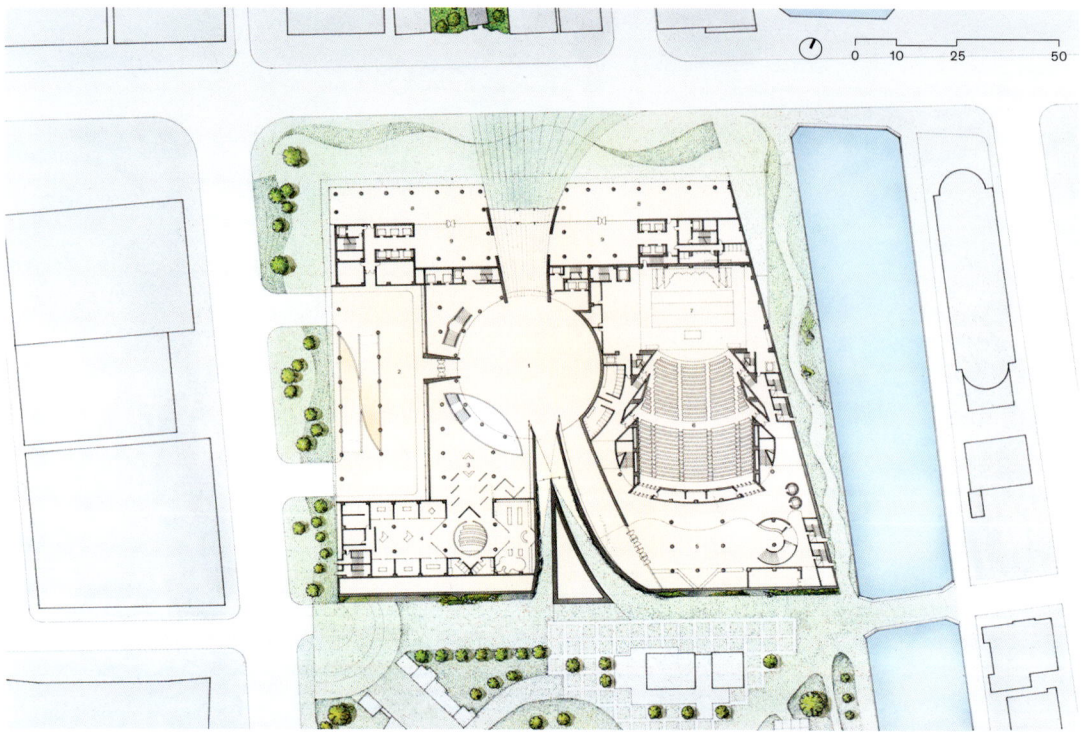

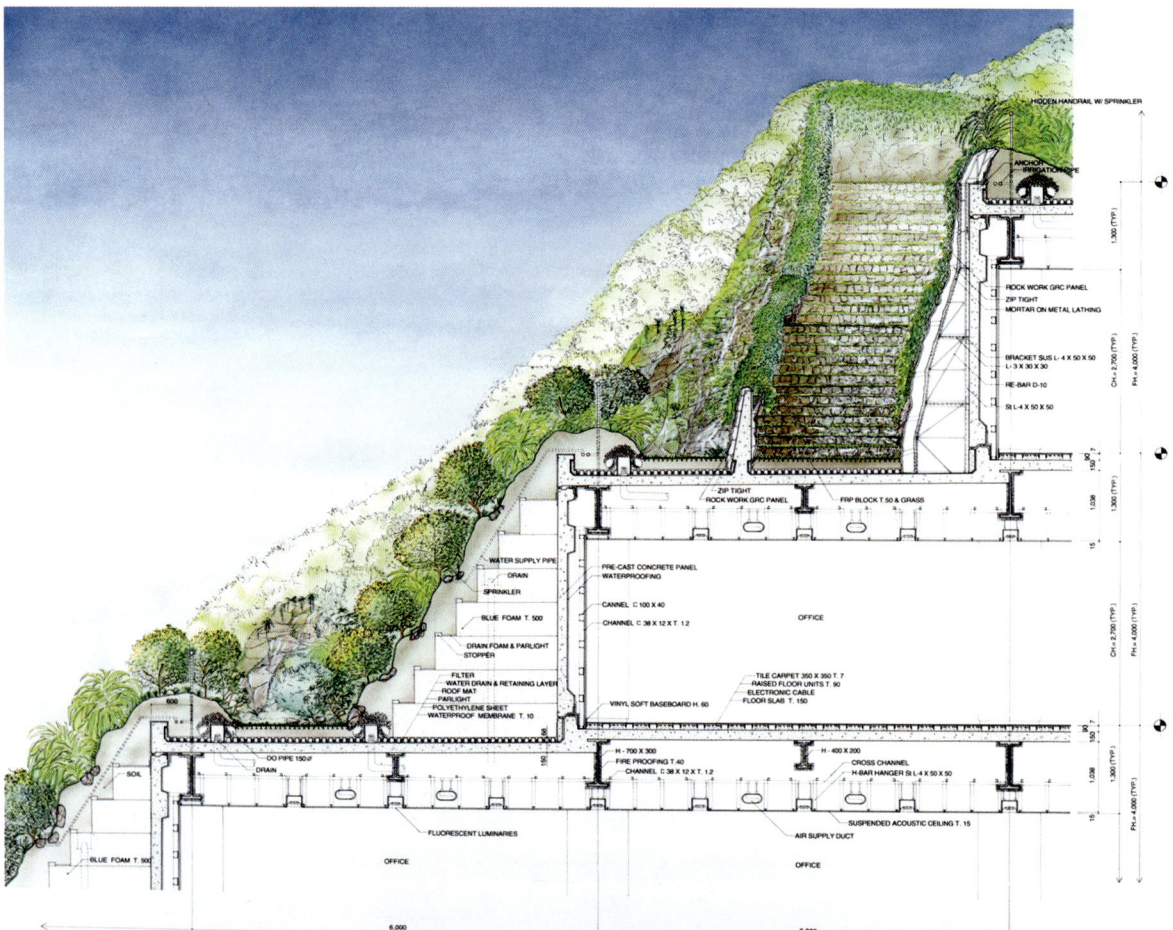

Fukuoka Prefectural International Hall

*Ambasz's shamanistic practice
is to make examples, to establish grounds
for collaboration, to assist the people
to trust their own insight when it comes
to moral and aesthetic judgment.*

Dean MacCannell

IDEOLOGICAL CASTLES

Dean MacCannell

[On the Burmese frontier, Kuki] houses were not so much built as knotted together, plaited, woven, embroidered and given a patina by long use. Those who lived in them were not overwhelmed by great blocks of unyielding stone; these were houses that reacted immediately and with great flexibility to their presence, their every movement. The house was, in fact, subject to the householder, whereas with us the opposite is the case. The village served the villagers as a coat of light elastic armor; they wore it as a European woman wears her hats. It was an object of personal adornment on a mammoth scale, and those who built it had been clever enough to preserve something of the spontaneity of natural growth. Leafage and the springing branch were combined, in short, with the exactions of a carefully planned layout. The inhabitants seemed protected in their nakedness by the fronded velvet of the partition walls and the curtain-fall of the palms. And when they went forth from their houses it was as if they had just stepped out of an enormous dressing gown of ostrich-feathers. Their houses were caskets lined with down [...] and their bodies the jewels within them.

Claude Lévi-Strauss, *Tristes Tropiques*

Daphne [is] transformed into a tree under the pressure of a pain from which she cannot flee. Isn't it true that the living being who has no possibility of escape suggests in its very form the presence of what one might call petrified pain? Doesn't what we do in the realm of stone suggest this? To the extent that we don't let it roll, but erect it, and make of it something fixed, isn't there in architecture itself a kind of actualized pain?

Jacques Lacan, *Seminar VII: The Ethics of Psychoanalysis*

Emilio Ambasz could easily have had Lacan's and Lévi-Strauss's words in mind when he wrote,

> [T]he ideal gesture would be to arrive at a plot of land which is so immensely fertile and welcoming that, slowly, the land would assume a shape—providing us with an abode. And within this abode—being such a magic space—it would never rain, nor would there ever be inclemencies. [...] In a perfect nature, we would not need houses.[1]

Ambasz's "ideal gesture" apparently does not originate in architecture. It originates in myth and in the unconscious. This much is already well established in the critical embrace of Ambasz's creative works.[2] *Myth* is the most frequently recurring trope in the writings about Emilio Ambasz and his work. For one who is steeped in the anthropological and critical study of myth, it is somewhat unexpected and touching to find this architecture submitting itself to an anthropological framing. Unexpected, because it does not appear to submit itself to any other framing. It eludes conventional architectural discourse; it is architecture-beyond-architecture. In a sympathetic essay, Ettore Sottsass remarks that Ambasz's buildings are "not architecture." They are more like "wagers."[3] The evocation of myth is touching, because technically Ambasz's work has no real need to be propped up with supernumerary theoretical discussion of its supposed mythic qualities. If anything, our understanding of myth will benefit more from its connection to Ambasz than vice versa. Anthropology should be honored that this exceptional architecture has reached out to it. In this essay, I will attempt to return the tribute.

It is Ambasz himself who authorizes mythic interpretation: "It has always been my deep belief that architecture and design are both myth-making acts."[4] And from his "Credo": "There is in all of us a deep need for ritual, for ceremony, procession, magical garments, and gestures."[5] Professional colleagues and critics have eagerly seized Ambasz's suggestion that his masterworks be understood in terms of their mythic significance. Peter Buchanan praises the architecture for renewing myth and "reenchanting" and "reintegrating" our "desiccated" and "fragmented" world.[6] Michael Sorkin names Arcadia as the mythic landscape of Ambasz's projects and thinking.[7]

After Saussure, Propp, Lévi-Strauss, and Barthes, our understanding of the operations of mythic thought is quite a bit advanced. We know that language is composed of phonic and grammatical oppositions and that myth derives its power from modeling itself on language. Myth is a viral copy of an original code. The building blocks of myth are bits of language that are detached from their historical coordinates and reassembled in persuasive ideological forms. The ideological function of myth is to allow its believers to live with social and cultural contradictions as if they do not exist, or if they do exist, to live with them as if they are neutral or meaningless.

Myth-Lite

When myth is summoned in writings by and about Ambasz it is not in terms that immediately resonate with anthropological or critical understanding. In the discussions surrounding Ambasz's projects, *myth* mainly references past epochs when humans were thought to have lived in closer harmony with nature, and were on more intimate terms with primordial mysteries. This critical strategy suggests that we should approach Ambasz's appeal to the mythic in design as a kind of "myth-lite." The Ambasz commentary only wants to go far enough to invoke the mystical, spiritual feel of myth; or the mythical surface of myth; or a sense of a mythic past; or a mythic oneness with nature. The commentary does not engage the structure of myth in any technical sense. An anthropological or critical consideration of myth would appear not to have a place in the discussion surrounding Ambasz's projects.

This tendency is evident in Peter Buchanan's essay, which pushes Ambasz's work quite hard in the direction of a New Age interpretation. According to Buchanan, Ambasz's projects "evoke certain urgent universals."[8] He praises Ambasz for reintroducing the mythic into architecture and design, claiming that it enables us to see beyond cause and effect and the bottom line, getting us back in touch with our primordial emotions.[9] He aligns Ambasz with the "two patron saints for the nineties, Carl Jung and Joseph Campbell," as well as the Gaia hypothesis, feminine archetypes, the Mother Goddess, and other theoretical elements that "encourage more nurturing, nonexploitative relationships between people, and between people and the planet."[10]

Without wishing to detract from the evident goodwill of Buchanan's and similar views, I am going to recommend against this version of myth as a way of understanding Ambasz's (and anyone else's) contribution. Elsewhere I have argued that this appeal to primitive innocence is not itself innocent.[11] We are supposed to identify with old or perhaps wiser societies because we too are good. But we have been expelled from Arcadia. In the New Age we understand ourselves as condemned to live in a too complex world, overwhelmed by thousands of meaningless, impersonal relationships, disconnected from family, from ethnic roots, and from an authentic sense of place. The New Age image we have created of so-called primitive peoples makes them the opposite of all this. They are simpler, more unified in their beliefs, more in touch with themselves and with nature, more spontaneous, intimate, altogether more magical, wonderful, and so on.

Thus, by a nasty theoretical sleight of hand, we see ourselves, not colonized native peoples, as the primary victims of our social and technological complexity. In this context, and against the expressed desires of those who advocate it, New Age mythology functions effectively as the enabling fantasy for the petty viciousness and actual violence that are the other marked features of the modern world. The suburban community undertakes to make itself a better place to live by mocking up a version of primitive homogeneity and uniformity of beliefs, by ridding itself of people who are different, and by surrounding each household unit with a patch of symbolic nature. This is an ideological perversion of the myth of the primitive. A gated suburb is a highly self-interested, motivated simulacrum of a primitive isolate, a fantasy of a fantasy. When it is examined in the light of a more rigorous understanding of myth, the embrace of putatively primitive beliefs and practices by hippies, New Agers, suburbanites, and other modern peoples is no less septic than forcing nonsustainable western technologies and extractive economies upon recent ex-primitives and peasants. The two sets of practices, which may want to see themselves opposed, are but the obverse and reverse of the hegemonic coin.

It seems unlikely, on the face of it, that this is the direction Ambasz's work is taking us, that his work is advancing some New Age hegemonic fantasy. Should we simply drop the mythic interpretation? I don't think so. Let me recommend that in place of a New Age mythology we reexamine Ambasz's projects in the light of more rigorous theories of myth.

Lévi-Strauss's "Science of the Concrete"

Lévi-Strauss's monumental lifework is a series of volumes that treats the meaning and function of one thousand American Indian myths culled from tribes covering the territory from the Bering Strait to Tierra del Fuego. What is remarkable about these myths, and Lévi-Strauss's handling of them, is that while they were independently collected from hundreds of quite distinct cultural groups, they can be seen as being composed from a restricted set of basic cultural oppositions, using a singular underlying transformational logic.

This method of understanding myth is not the same as that of Jung. It makes no appeal to universal archetypes. In fact, Lévi-Strauss goes in an opposite direction. Where there appears to be a common mythic theme shared by different peoples, Lévi-Strauss shows how each culture bends and twists it, turns it inside out, transforms it into its opposite, and otherwise reshapes it so it works in the immediate cultural context—not as something borrowed, but as a crucial stitch in the local cultural fabric.

The anteater appears to be an archetype in South American myth, but in one group he is found resolving the tensions produced by an incest prohibition, while in another he shows the people the way to a more habitable territory. There may be figures that appear to be universal, or nearly so. But no figure is universal in its meaning and function—that is, there are no mythic archetypes. There are only more or less effective strategies for trying to understand the world and our place in it. According to Lévi-Strauss, myth is protoscience. Does this make it less heartfelt or emotionally satisfying? No. It serves to mark and to honor intense creativity, our most human attribute.

Lévi-Strauss discerns in the myths he studied a speculative thinking of the sort our ancestors must have needed to create the great inventions of the Neolithic revolution: writing, kinship, universalistic religions, plant and animal breeding, ceramics, calendars, and, relevant to the topic at hand, architecture. Native American myths provide an integrated framework for our most durable form of theoretical thought: the systematic effort to fit everything in the world together in a way that makes sense. So-called primitive societies must contain and somehow resolve primary and secondary structural oppositions: for example, nature versus culture, or the raw versus the cooked. Myth inserts itself between these oppositions and builds structural sets out of the "debris of events"[12] or "ideological castles" from the remains of discourse.[13] Even though myth restricts itself to existing symbolic constructs, restless "mythical reflection can reach brilliant unforeseen results on the intellectual plane."[14] Myth orders events in a search for meaning. Science orders ideas (meaning) in order to produce events.[15]

Myth, according to Lévi-Strauss, permits people to live with the structural contradictions peculiar to their cultural arrangements, as if they were not contradictions. It operates at the collective level and also at the psychic level. Or, as Lacan put it, "Myth is always a signifying system or scheme, if you like, which is articulated so as to support the antinomies of certain psychic relations."[16]

Ambasz's "Science of the Concrete"
Ambasz has not tried to discourage an interpretation of his work in terms of "myth-lite" à la Jung, Campbell, Buchanan, and New Age clichés. Indeed, there can be little question that he has actively sought to make his work accessible in these terms. But there is another side to Ambasz's self-analysis, at once darker and more reflective, like an obsidian mirror, that suggests a harder stance on myth and politics. This alternate line of thought about myth and mythmaking is summarized in his essay "Coda: A Pre-Design Condition."[17] "Coda" begins with some (insider?) observations of the events of May, Paris, 1968. On the first page, Ambasz expresses disappointment that the revolution succumbed to its own myth of immediacy. He sadly observes that nothing remains except the "paper barricades." The rest of the essay is his effort to recoup from the failed revolution his youthful enthusiasm for social change. By the second page of this fascinating text, Ambasz is engaging in speculation about structure that is either heavily indebted to Lévi-Strauss or a dramatic case of parallel invention.

> Man creates structures that are the meeting grounds between two incommensurable realms. […] A structure […] is designed to reconcile the discordant interface. […] Particular instabilities are likely to follow its attempted resolution of the conflicts between the two realms.

> A structure may act as a pattern breaker; it may dissolve established correspondences between thought and matter and introduce new ones; it may induce a physical rearrangement in the infrastructure and an epistemological renewal.[18]

Lévi-Strauss sought to explain how mythic thinking, constrained to work with a bricolage of existing cultural elements, could lead to the intellectual creativity that produced the inventions of the Neolithic revolution. How did our primitive ancestors, in possession of nothing more than mythic thought, no Galilean insight, no Cartesian method, no experimental research design, and no mathematics, reach "brilliant unforeseen results on the intellectual plane"? Nowhere does Lévi-Strauss describe the likely process any more succinctly than Ambasz does here.

Ambasz consciously locates his architecture (indeed, all architecture) in the space of myth, not "myth-lite," but the mythic operations that are located between two discordant realms. In the following sections I offer a listing of the fundamental oppositions, the discordant realms underlying Ambasz's architectural practice.

Primitive / Modern

Of the series of oppositions that Ambasz interrogates through his work, the most obvious for an anthropological reading is primitive/modern. It is also the most reflexively self-conscious opposition in his architectural idiom. It is what miscued some of his admirers into thinking his work can be contained by a New Age reading.

Adroit critical handling of the primitive/modern opposition requires us to scrape it clean of its encrustation of sophistry about crystals, feathers, magic mushrooms, and the mother goddess. Ambasz's projects embody the essence of the Primitive. Yet, in their presence we do not get a whiff of someone "going native" or "playing Indian." In fact, if there was ever a contemporary artist whose work is free of literal references to the primitive and primitivism it would be Emilio Ambasz. He returns us to a primitive state without returning us to a primitive state. The question for the critic is, How does Ambasz work this peculiar magic?

Alessandro Mendini tells us that Ambasz's work "is born from an obsessive search for primary principles."[19] It does not, however, reproduce the primitive state and hold it up as an example. It restages the fundamental problematics that we, as humans, have had to deal with from the beginning. Ambasz's creative response is on the same footing as that of our primitive ancestors', but time has passed and he must address the cultural crudation that has grown up around the original solutions, in addition to proposing new solutions. This is why Ettore Sottsass can say of his work with such evident truth that it looks like "a thousand hours after the cataclysm" when humankind has found a way to "restructure some form of mental and physical survival," to "build a stairway leading to who knows where."[20]

Michael Sorkin astutely retraces a trajectory that was first calculated by Claude Lévi-Strauss, transforming primitive/modern into minimal/elaborated. Sorkin remarks with a Rousseauistic twist that Ambasz's work embodies a "pared-down integrity of innocence" without "irony or suspicion."[21] And, "it is not about being simple. It is about being concise."[22] We might add that it is a minimalism without a trace of Puritanism. Ambasz does not elevate the primitive over the modern, but he treats the two as a technical opposition that provides an opening for rigorous interrogation of what is basic and true about human emotions, relationships, and creativity.

Nature/Culture

We have before us skyscrapers of trees, mountains of glass, buildings that are floating islands, and a train station that is a desert oasis. Ambasz comments:

> I seek rhetorically to eliminate architecture as a culturally conditioned process and return to the primeval notion of dwelling. I seek to develop an architectural vocabulary outside the canonical tradition of architecture. It is an architecture that is both here and not here.[23]

Ryuichi Sakamoto suggests the oeuvre taken as a whole is a metaphoric scale model of planet Earth. It reminds him of "a gigantic space station endowed with desert and savanna areas where ten million people dwell."[24] In this new architecture, nature/culture carries no weight as an opposition, but it carries ultimate weight as a pair. Sorkin elaborates in landscape terms: "About the work is an abiding sense of the ecological, nature's own system of decorum. The sort of reticence about grade that the work persistently displays is part of this, a deference to mother nature."[25] And with regard to the other member of the nature/culture pair, Sakamoto tells us the work is not about just one building after another, but it "always bears in mind an entire culture."[26] Sakamoto goes on to state that the work promises to "reestablish a pact of reconciliation with nature."

Tadao Ando says Ambasz uses "nature on a massive scale," presenting "the entire environment as a constellation from which architecture draws its essential being." And, "there is [...] no prior example of nature governing architectural creation with such power and haunting seduction."[27] To this we could add that there is no prior example of architecture governing nature with such power and seduction. In the Houston Center Plaza, the cubic geometricism of the trees below and the dance of vapors above constitutes a new form of the nature/culture opposition. Ambasz takes pains to create places that are neither God-given nor man-made, but precisely suspended between the two realms.

There is a critical consensus that Ambasz's architecture defers to Mother Nature. It embodies an "ethic of growth" that is "decorous," "reticent," "innocent," given to "gentle gradients," "nurturing," "nonviolent," and "nonaggressive." Can't we just call it "green architecture"? Yes, but it is a unique kind of green architecture, one that does not permit itself to succumb to crude do-it-yourself aesthetics. Ambasz's heroic efforts to hold himself accountable to principles of ecology and cultural history has the mythic potential for tragedy. There is always something sorrowful here, because the work already stands for the loss that accompanied the millions of pieces built by others in the progressive desolation of Earth.

Front / Back, Surface / Depth
It has been said of postmodern architecture that it is all surface and no depth. Fredric Jameson and others have taken this to be the organizing metaphor for postmodern culture in general: there is no depth of understanding, feeling, or historical sensitivity. Most of the writings about Ambasz comment that his work does not contribute to the postmodern aesthetic. To distinguish it from postmodern exer-

Houston Center Plaza, Houston, Texas, USA, 1982

cises, Buchanan states that Ambasz's work has "little in the way of facades."[28] Buchanan must mean *facade* in the sense of false fronts, or overly decorated fronts, not in the original, structural sense of the face or front of a building. All architecture has some kind of facade to divide inside from outside and front from back. It would be better to say that the facades of Ambasz's buildings work a mythic inversion of front and back and renew our involvement in the passage from inside to outside and out to in. In his interiors he favors walls made of light. And at the Casa de Retiro Espiritual, there are walls without an inside. No one has theorized this better than Ambasz himself:

> You always have a sense that behind the walls of these projects are absent presences or present absences. The notion of that which is in front of you and what happens behind the wall has always appealed to me. There is a certain anima or spirit behind the wall. The berms act as symbolic fortresses through which you pass to discover a terribly benign giant. [...] I have no wish to provoke anguish.[29]

In execution it gets even more interesting. In a characteristic gesture at the Banque Bruxelles Lambert, he installs the building's facade inside, in the lobby. He speaks directly to the complexities of psychic desire for intimacy. The designs promise us access to a mysterious interior and at the same time allow us a way out. Like an understanding illicit lover, they are not cloying; they never trap us in an unwanted intimacy.

Ambasz's handling of inside/outside disrupts the taken-for-granted alignments of architecture's conventional antinomies. Tadao Ando complains that, in modern times, "nature came to be treated as a mere visual accent, a mere aspect of landscape, a subject of adornment, and too often it [is] relegated to the margin of the site."[30] Not with Ambasz.

	Inside	Outside
Culture	MODERNIST AESTHETICS	AMBASZ
Nature	AMBASZ	MODERNIST AESTHETICS

Modernist aesthetics took pains to elevate culture over nature, and postmodernism elevated surface over depth. Ambasz strives to hold the two poles of both oppositions in a relation of perfect equality. And he allows the balanced oppositions to oscillate around each other, functioning as mythic operators, producing "brilliant unforeseen results."

Sacred/Profane

A few of Ambasz's projects have called on him to produce sacred sites or to deal directly with spiritual themes. These include the Casa de Retiro Espiritual in Spain, the convent in Singapore, and the small belowground chapel for the agricultural workers' collective in California. While the numbers of specifically sacred buildings are small, the commentary on Ambasz's work as a whole very often invokes its alignment with sacred values. Buchanan generically calls the work "sacramental." Here again, Ambasz lends his own authority to this critical turn, commenting, "We should allow designers to act openly as once the shamans of 'old' societies, perhaps wiser societies, acted."[31]

It is commonplace to the point of being taken for granted that contemporary life is no longer infused with a strong sense of the sacred. Buchanan says that our lives are "drained of the radiantly sacred [...] dried up and fallen apart."[32] This belief that we have lost our ability to believe lends a certain urgency to comments about the sacred quality of Ambasz's architecture. The claim is that he is helping us find our lost spirituality and making us whole again. I propose that there is something important going on here, but this is not it. If a spiritual sense of the sacred really drained out of modern life, the opposition sacred/profane would merely coincide with traditional/modern, and a renewed sense of the sacred would be a simple reprise of traditional forms, a kind of retro spirituality à la the Family Values Movement. This does not feel like Ambasz on the face of it. Something more complex is happening.

Technically, a sense of the sacred, the awe that the sacred inspires, requires a special kind of writing on objects and events that simultaneously marks them as embodying ultimate values and separates them from secular, profane objects. Durkheim said that the sacred is the original "value added."[33] Sacred writing comes from the social group, but the group must always convince itself that it comes from without. Among the oldest forms of this kind of writing that designates parts of the world, the law, and experience as *sacred* are the hallowed *names* that are given to places and individuals and protected by ritual.

The notion that the sacred has departed modern life is simply untrue. This is another of those myths of modernity that are necessary to secure the position of ruling interests in our modern societies. (See "Myth Today" below.) If we the people are convinced that we have become separated from our own deep spirituality, then corporations, politicians, and other phony casuists can profit by selling their diluted versions of morality and spirituality back to us. This is how it became possible for abortion, which has no political significance whatsoever, to become a major political issue. ("At last, an issue you can believe in.")

While any body would serve, the easiest proof of the continued deep resonance of sacred values in contemporary societies is the ideal feminine body nicely shrink-wrapped as it is in moral lustrations. Our sacred bodies can be tainted by even the slightest odors. Our garments—indeed our very being—are defined by the smallest stain. Our teeth must be at least three shades whiter or we shouldn't dare to open our mouths. The world that we live in continues to be much more religiously defined than we admit. As Goffman once famously commented, "Many gods have been done away with, but the individual [...] stubbornly remains as a deity of considerable importance."[34] Only now the commandments are brought down from the mountain by Proctor and Gamble and the Republican National Committee. Every modern person, especially the pitchmen, seducers, connivers, and big liars among us, is surrounded by a wall of morality that protects the sacredness and dignity of the image they want to claim for themselves. None of this would be possible if our world was devoid of sacred meaning.

    I disagree with the comments to the effect that Ambasz is reintroducing a sense of the sacred back into architecture and contemporary life. The sacred has been subject to theft and manipulation, but it has not departed the modern social world. Power continues to depend on controlling sacred values as it always has. The strategy for taking the sacred away from the people is fully formulated. It requires two steps: (1) convince the people that their ordinary, everyday lives are drained of all spirituality, devoid of heightened meanings, and (2) offer the people compensation for their lost spirituality in the form of commercialized entertainments packaged as spectacle, blockbuster movies, show trials, celebrity weddings, beauty pageants, and so on. The pretense that nothing significantly meaningful or of spiritual value can occur in the ordinary, everyday life of an average person runs precisely counter to the insights of the human sciences. According to Marx, Freud, and Goffman, everyday human relationships are the basis for our most intense psychic dramas. Conflicts, the formation of identity and character, and the everyday relation to production form the engine of history.

    By decrying the "loss of the sacred," the most passionate critics of contemporary social arrangements hand victory to their enemies on a silver platter before any discussion or debate even begins. It is only to the extent that we actually believe our everyday lives are drained of the sacred and the consequential that we can have our humanity stolen from us and a crappy version of it sold back to us. It is only when we think our lives are meaningless that we turn to corporate authority to tell us what recreational vehicle is needed to commune with nature, who is the current god of popular culture, who has a perfect body?

In this context, all serious architecture can be seen as a form of sacred writing. It is not possible to rearrange a pile of stones without invoking the sacred. The issue here, for Ambasz in particular and architecture in general, has to do with the placement and definition of the line between the sacred and the profane, and the ownership of the means of production of the sacred. Does architecture express the dignity and meaning of the everyday, or does it seek to impress us with the putative insignificance of our ordinary lives? Ambasz does not bring back a sense of the sacred. Rather, he moves the battle lines between the sacred and the profane as they have been defined in the recent culture wars, or, more properly, in the current corporate war against culture. This is clearly enunciated in his "Credo":

> In my architecture I am interested in the rituals and ceremonies of the twenty-four-hour day. I am not interested in the rituals for the long voyage. [...W]hat a tragedy to discover that for the sake of those long-term dreams, we have sacrificed our daily lives. No, I am interested in daily rituals: like the ritual of sitting in a courtyard slightly protected from both the view of your neighbors and the wind—looking up at the stars. Dealing architecturally with that type of situation attracts me.[35]

Ambasz's architecture informs us that the raw materials for constructing an opposition between the sacred and profane are more or less evenly distributed between peoples, localities, and historical moments. He takes the humblest materials, a quiet moment in a courtyard, a depression in the earth, a small tree that appears to have volunteered to grow on a stairway, and transforms them into sacred objects and events. Sottsass poetically remarks:

> I see [Emilio Ambasz] as an imaginative and illuminated young man, resembling some old Chinese priest who, first for months, and then for years, uses ancient techniques to polish the surface of a great disk of green jade which will allow him, perhaps, to penetrate beyond the daily, or better still, which will allow him, even for just an instant, to place daily existence in some exalted architectural domain.[36]

What is Ambasz up against? Corporate dream merchants adept at convincing the people that they are not competent to comprehend and shape materials to satisfy their spiritual needs. Ambasz's shamanistic practice is to make examples, to establish grounds for collaboration, to assist the people to trust their own insight when it comes to moral and aesthetic judgment. This is the way Ambasz understands it:

> We should allow designers to act openly as once the shamans of "old" societies, or perhaps wiser societies, acted when they were more attuned to the human imperatives of dealing with passions, emotions, wonder.[37]

There is still a great deal of work to be done before everyone can see that we are capable of creating our own meaningful moments, that the women in the street are more beautiful than the women in the beauty pageant, that the important building doesn't have to be a phallic monstrosity.

Solid / Liquid / Gas

Since Aristotle, these categories have appeared on every list of concepts that permits us to think about essential differences in the form of substance, and of substantiality itself. Along with time, space, class, and number, they are among our most precious instruments of thought. Architecture, by convention and by code, is inscribed mainly within the category of the solid. Insofar as concepts require representation, buildings, along with their favored material, stone, are among the most privileged signs of solidity.

Emilio Ambasz's buildings do not sit proudly on the surface and appear to be gloating about their solidity. Instead, they hang in the air, or they are rhizomatically connected to everything that is belowground. They are an affront to the kind of architecture that asserts its substantiality by mimicking the cube. Ambasz has created a distinctive architectural vocabulary that delights in its alignment with fluidity and insubstantiality. This has been pointed out by Ettore Sottsass, who remarks, "Such a quest for a constant state of fluidity, such a perception of existence as an ever-changing process pervades everything that Emilio has ever designed."[38] The play of forms that are more proper to a liquid than to a solid animates even the most prosaic details of his construction. Ambasz explains that before he buries a mobile home, he calls in a swimming pool contractor to blow gunite around the outside of the structure that will be under the earth: "I can take a normal house, a Levitt house, or even two trailers perpendicular to each other and cover them partially with earth. [...] For projects of this sort, I employ people who build swimming pools. [...] Waterproofing is the last step before extra earth is brought in and mounded against the walls."[39] What is this thing except a pool turned inside out and made into an underground dwelling?

Here, for reasons that should be evident enough in a moment, I feel a strong need to go back to Lévi-Strauss's study of myth. Few of the myths Lévi-Strauss analyzes articulate the origins of house-building or architecture. The following Mataco myth may be the only one. It was collected by Metraux in the 1930s in northern Argentina about

750 miles from where Ambasz was born. The Mataco people were nomadic hunters and gatherers ranging in the Gran Chaco region of Argentina, near disputed borders with Paraguay and Bolivia. According to Mataco mythology, in the beginning there were no houses and no knowledge of methods needed to build them. We can pick up the story the way Lévi-Strauss retells it:

> [A]n Indian one day heard a noise at the bottom of a lagoon. He dived in and saw people building a large house under the direction of a master carpenter. This carpenter was none other than the skate [sting ray], who taught the Indian the art of house building. [. . .] T]he skate was the first carpenter.[40]

I do not know if Ambasz has heard this story. I would like to know. No other story slips so easily into dialogue with Ambasz's imaginary. The theme of the submerged house, of drawing inspiration from nature, and from unlikely sources in nature: everything in the Mataco myth has found its expression in Ambasz's work.

If there are underlying principles to Ambasz's architectural practice, one of them would certainly be that in the solid/fluid opposition, architecture should align itself more with the fluid and less with the solid. The Schlumberger Research Laboratory calls for a prairie with some "natural" features: hills, valleys, a stream, and a lake. The earth of the prairie is strategically cut to accommodate offices and laboratories, half-underground. The individual offices of the scientists are to be parked here and there under the landscape. Each office unit is self-contained and mobile, allowing the scientists to select their neighbors according to their preferences and collaborations of the moment, and to select their views according to their inspirational needs. The flexibility of this arrangement would seem to respond directly to Lacan's dictum (in Seminar IV) that the most basic human desire is to be somewhere else. It also answers Lévi-Strauss's invidious comparison of Western architecture to the Kuki houses he describes in the passage at the beginning of this essay: "Those who lived in them were not overwhelmed by great blocks of unyielding stone; these were houses that reacted immediately and with great flexibility to their presence, their every movement. The house was, in fact, subject to the householder, whereas with us the opposite is the case."[41]

The same principle is found in the proposal for the Seville Exposition. The pavilions are floated on artificial lagoons. The entire design acknowledges the impermanence of international expositions. It is made to be recycled into a university campus. And when each of the pavilions becomes a university department, these can be flexibly arranged and rearranged into colleges in response to changing patterns of interdisciplinary collaboration. Of course, not everyone would be comfortable with so much flexibility. Resistance to this plan would nicely reveal irresponsibility in the face of choice, and hypocrisy in the embrace of freedom.

This inversion of the solid/fluid opposition is found in the program of many works, and it is also evident in the aesthetic elements of virtually every design. Exterior surfaces woven from living trees recall the Yuki houses described by Lévi-Strauss. The original design for the Fukuoka Prefectural International Hall called for a folded and inclined lake, as if not even the fluid is flexible enough for Ambasz. In summary, I will again paraphrase Sottsass: "Emilio's buildings are not objects. They are not architecture. They invoke the presence of architecture. They are more like wagers."[42]

Ambasz's architectural practice contains the outlines of an alternative modern mythology—not "myth-lite." Fumihiko Maki comments that Ambasz has created something new through "violent oppositions."[43] This remark appears to stand in direct contradiction to Sorkin's and others' assessment that the works are "reticent," "innocent," "gentle," "nonaggressive," and "nonviolent." Can both claims (for violence and for nonviolence) be correct? I think so, at the level of mythic transformation. What Ambasz creates is nonviolent. But it is necessary for him to do violence to conventional oppositions and hierarchies to arrive at his distinctively nonviolent architectural result. Thus his work stands not just as architecture but as strong critique and comment on the state of the world we live in, set up as it is to underwrite all forms of violence. The particular revisions Ambasz is making to the antinomies that underlay everyday life is the reason his work is inclusive. It is compelling not just to architects and artists but to anyone who might stumble across it. This is not because it invokes universal mythic archetypes. It is because it enters and works the gaps in human consciousness that have been created by myth.

Myth Today

Roland Barthes extended the study of myth to cultural arrangements under late capitalism. According to Barthes, modern myth is not necessarily found in formulaic narratives or in other material that announces itself as myth. For example, his essay on "Ornamental Cookery" focuses on the presentation of food in the French magazine *Elle*.

> Glazing, in *Elle,* serves as a background for unbridled beautification: chiseled mushrooms, punctuation of cherries, motifs of carved lemon, shavings of truffle, silver pastilles, arabesques of glacé fruit. [...]t is because *Elle* is addressed to a genuinely working-class public that it is very careful not to take for granted that cooking must be economical.

> [... O]rnamental cookery is indeed supported by a wholly mythical economics. This is an openly dreamlike cookery, as proved in fact by the photographs in *Elle,* which never show the dishes except from a high angle, as objects at once near and inaccessible, whose consumption can perfectly well be accomplished simply by looking.[44]

According to Barthes, myth today is not so much built from the debris of discourse as it is embedded in it. We do not relate our myths so much as we bear them on our persons and in our belongings, in figures of speech, clothing styles, automobile design, advertising formulas, representational strategies, motion picture clichés, retro and nostalgic architectural styles, decoration, toys, and gadgets.

While Lévi-Strauss's and Barthes's theories of myth were given to us in very different forms (Barthes's was in several short essays, not several heavy books), they are markedly similar in their essentials. Modern myths, no less than "primitive" myths, are ideological castles built from scraps of discourse, exactly as Lévi-Strauss suggested. But Barthes's words about myth work are much stronger than Lévi-Strauss's. Barthes says that myth is "parasitic," that it "appropriates" and "nourishes itself" on history, and "distorts" language for its own ends.

What kind of work does myth do in late capitalist society? Again, it is precisely the same as that posited by Lévi-Strauss for American Indians. Our myths permit us to live our social contradictions as if they are not contradictions. The difference between primitive and modern myths are the actual historical contradictions that the myth work covers up and supports. How do we reconcile the democratic principle of human equality with the racism that is still rampant in democracies, or with the enormous difference in life's possibilities for the wealthy and those for the impoverished? Modern myth glosses over these differences that have specific historical origins by pretending they are not historical but a part of the natural order of things. Thus, we have a thousand variations on the myth that free enterprise is human nature; or the myth that blacks are naturally inferior to whites, but are suited to certain types of work, good at sports, and capable of loyalty. Myth appropriates language to suppress history. It transforms conservative values into "facts." It enables a kind of right politics that appears to operate beyond merely political considerations by wrapping an elevated class perspective in pretentiously apolitical, universalistic rhetoric. Conservative mytho-politics hides its specific agendas behind vapid generalities: "All people have the right to be free from government regulation of their authentic human potential."

While the fundamentals of the theory of myth remain remarkably stable and constant across the primitive/modern divide, the moral and ethical implications are entirely different. Why? Because the histories are incommensurate. No primitive society ever had to deal with the kind of human inequality that we moderns face on a daily basis. When Montaigne sent his servant to interview a chief who had been brought to Europe from the recently discovered Americas, he inquired first of all about inequality, or the material consequences of status difference. Did chiefs dress any differently, or eat better food, or live in larger or more comfortable houses? No. Were there any differences then? Yes. The chief proudly proclaimed two differences. When the group was on the move through the forest, his subjects would pull back the foliage to make it easier for him to pass. And the second difference? When his people went into battle against a mortal enemy, it was the chief who was privileged to lead the charge.

Evidently we're not in Arcadia anymore.

How do we reconcile this critique of myth under capitalism with an architecture that simultaneously espouses democratic values and wants to think of itself as myth work?

Barthes does not directly suggest, though he strongly implies, that the gap between rich and poor under extreme capitalism is great and unbridgeable. The middle class might be pivotal, but there is simply too much ideological terrain to bulldoze, and our collective understanding of history is too weak. Accordingly, the "middle class" spends its considerable wealth and energy on a compensatory myth with which it is endlessly preoccupied. Whatever revolutionary potential the middle class might possess has been completely neutralized by its distinctive consciousness built from the droppings of mythic discourse: gleaming domestic surfaces indicate social standing; the appearance of compassion is equal to the reality of justice; the natural destiny of woman is to be a wife and mother; the wealthy work harder for their money than the poor; social and other kinds of friction can be made to disappear from our lives; if someone fails it is their own fault; brilliant economic success is the result of a good business plan; shopping is patriotic; it is possible to insulate oneself from socially undesirable elements by living in a gated community; with sufficient social and other prophylaxes sex is completely sanitary; and so on. According to Barthes, the bourgeoisie is historically nothing except what has been pasted together from these and similar mythic beliefs. He calls the bourgeoisie a "joint stock company" and remarks, "Ideologically, all that is not bourgeois is obliged to borrow from the bourgeoisie."[45]

Where does Ambasz stand on the matter of modern mythology having been fully co-opted in the service of reinforcing existing social hierarchies? For example, myth today as nothing more than the ideological support for the singular opposition in late capitalist societies: that between wealth and poverty, between surplus and lack. This is more difficult for Ambasz, and for us, than his alignment with Lévi-Strauss's theory of myth. He is clearly ambivalent on the matter of the politics of his mythic experimentation. Sometimes he denies that his works have any political significance: "I want to be a fabulist, not an ideologist."[46] At other times, he deals directly, and often bravely, with political questions and theory. He prefaces these moments with a quick move to dissociate his work from traditional class-based critiques.

> [T]he traditional agent of change, the class subject, is spottily assumed by a lucid and sensitive few who place themselves on the periphery of the system. They may attempt to establish alliances with those who are dispossessed or deprived of their rights, but these are now outside channels of production and wield no power; and the alienated ones who do take part in production can no longer recognize themselves as a social class.[47]

I read this as fully acknowledging the effective political neutralization of all class perspectives except, of course, the highest one. In other words, according to Ambasz, myth won, not Barthes's exposé of its inner workings. This would be a bitter pill for both the old and new Left to swallow. But the Far Right should not begin their final victory dance. None of the problems that gave us Left and Right have gone away. If anything, they have intensified. Ambasz does not stop with his critique of leftist rhetoric, he immediately tries to shift the struggle for the soul of humanity onto a different ground.

> Formal organizations [...] legitimize social relations, the ownership of resources, etc. [... T]he structures that man designs, the myths he creates, are temporary crystallizations in the perennial state of dialectical transaction between the fears and desires underlying the individual's aspirations and the assembled forces of the natural and sociocultural worlds.[48]

In short, the historical framing of human aspiration in terms of social class position has only driven revolutionary sentiments deeper underground. The class positions that have been proffered are not attractive choices: to be an oppressor, or to desire to be an oppressor. Ambasz wants to start with a clean sheet on these matters, which makes his architecture historically interesting indeed.

> [E]ven though the old [class] distinctions have been blurred, conflicts, albeit of a different nature, still exist between infrastructure and superstructure. The strategy that is called for is the design of structures which, mediating between the two levels, attempt to resolve their new kinds of conflict, and to cause in both changes that are wanted.

---

1 Emilio Ambasz, "Fragments from My Credo," in *Emilio Ambasz Inventions: The Reality of the Ideal* (New York: Rizzoli, 1992), 55.
2 Michael Sorkin, "Et in Arcadia Emilio," in *Emilio Ambasz: The Poetics of the Pragmatic: Architecture, Exhibit, Industrial and Graphic Design* (New York: Rizzoli, 1988).
3 Ettore Sottsass, translation of catalog essay in *Architettura naturale: Emilio Ambasz progetti e oggetti* (Venice: Electa, 1995), xxii.
4 Emilio Ambasz, "I Ask Myself," in *Emilio Ambasz: The Poetics of the Pragmatic*, 24 (see note 2).
5 Emilio Ambasz, "Fragments from My Credo," 53 (see note 1).
6 Peter Buchanan, "Emilio Ambasz: The Relevance of Resonant Ritual," in *Emilio Ambasz Inventions*, 27–29 (see note 1).
7 Sorkin, "Et in Arcadia Emilio," 19 (see note 2).
8 Buchanan, "Emilio Ambasz: The Relevance of Resonant Ritual," 15 (see note 6).
9 Ibid., 29.
10 Ibid.
11 Dean MacCannell, *Empty Meeting Grounds: The Tourist Papers* (London: Routledge, 1992), 292ff.
12 Claude Lévi-Strauss, *The Savage Mind* (Chicago: University of Chicago Press, 1966), 22.
13 Ibid., 21.
14 Ibid., 17.
15 Ibid., 22.
16 Jacques Lacan, *Seminar VII: The Ethics of Psychoanalysis* (New York: W. W. Norton & Company, 1992), 143.
17 Emilio Ambasz, "Coda: A Pre-Design Condition," in *Emilio Ambasz: The Poetics of the Pragmatic*, 49ff (see note 2).
18 Ibid., 50.
19 Alessandro Mendini, catalog essay in *Emilio Ambasz: The Poetics of the Pragmatic*, 16 (see note 2).
20 Ettore Sottsass, catalog essay in *Emilio Ambasz: The Poetics of the Pragmatic*, 10 (see note 2).
21 Sorkin, "Et in Arcadia Emilio," 17, 21 (see note 2).
22 Ibid., 21.
23 Ambasz, "I Ask Myself," 28 (see note 2).
24 Ryuichi Sakamoto, "Return of the Entire Humankind to Earth," in *Emilio Ambasz Inventions*, 11 (see note 1).
25 Sorkin, "Et in Arcadia Emilio," 22 (see note 2).
26 Sakamoto, "Return of the Entire Humankind to Earth," 9 (see note 1).
27 Tadao Ando, "Amplitude's Promise Fulfilled," in *Emilio Ambasz Inventions*, 43 (see note 1).
28 Buchanan, "Emilio Ambasz: The Relevance of the Resonant Ritual," 17 (see note 6).
29 Ambasz, "I Ask Myself," 27–28 (see note 2).
30 Ando, "Amplitude's Promise Fulfilled," 41 (see note 1).
31 Emilio Ambasz, "Architettura radicale," in *Emilio Ambasz Inventions*, 76 (see note 1).

What these structures can and ought to be cannot be known in advance of their existence. However, as we have seen, in an advanced industrial society the conditions do not exist which permit the emergence of structures that may bring about significant change. We are, so to speak, in a pre-design situation.[49]

Are we, so to speak, in a perpetually prerevolutionary situation? I can think of no more frustrating place for humanity to end up. But this is also the exact point of theoretical and critical impasse. Where do we go from here? Toward the prospect of a revolutionary bourgeoisie? Or straight to cyber-nuclear-fascism?

This is not just a problem for the Left. The Right has its work cut out for it as well. At this moment, the Right has positioned humanity between traumatic responses to staged events and nonresponses to actual cataclysms. If anything, this is a tougher space to work one's way out of than the impasses on the Left. There is no middle way out of this dilemma. Or there may be, but the so-called middle way is always further to the right than the conservative way. Recall that Hitler and the Nazis extolled the virtues of German National Socialism as the sensible "middle ground" between the "excesses" of American capitalism and Soviet communism. And they knew excess.

It is important to continue to examine the role of myth in society today. Clearly, Barthes's version of myth is not the same as Ambasz's. Barthes was incensed by myth's capacity to neutralize the historic role of the working class and to deny its contribution to our general well-being. He was equally incensed by myth's capacity to disguise the cruel, self-interested manipulation of all of our lives by the controlling class. In short, Barthes was infuriated by myth's capacity to drain all historical meaning out of the structural opposition that sustains capital. Ambasz's strategy is to step around social class and interrogate the superstructure that supports it. Ambasz does not place one member of an oppositional pair over the other, or try to drain all the meaning out of both. Rather, he seeks to create a new balance and to infuse both sides with sumptuousness. Even the member that had been privileged by previous cultural movements is enriched in his hands. It is a matter of aesthetics, but it is also political.

Ambasz has selected the most consequential terms possible to frame his own work. He has thrust himself and his critics into the historical game with the highest stakes. ("His architecture is more like a wager.") If it is just a stronger kind of glue to hold an ultimately unsustainable class structure together, it will only serve to make the eventual explosion that much more destructive. If, on the other hand, it is a brilliant engine for generating alternative human arrangements and new models of governance, there are grounds for hope.

32 Buchanan, "Emilio Ambasz: The Relevance of Resonant Ritual," 29 (see note 6).
33 Emile Durkheim, *The Elementary Forms of the Religious Life* (New York: The Free Press, 1965), 363.
34 Erving Goffman, *Interaction Ritual: Essays in Face-to-Face Behavior* (New York: Anchor Doubleday, 1967), 95.
35 Ambasz, "Fragments from My Credo," 53 (see note 1).
36 Sottsass, translation of catalog essay in *Architettura naturale*, xxiii (see note 3).
37 Ambasz, "Architettura radicale," 76 (see note 1).
38 Sottsass, translation of catalog essay in *Architettura naturale*, xxii (see note 3).
39 Ambasz, "I Ask Myself," 27 (see note 2).
40 Claude Lévi-Strauss, *The Naked Man: Introduction to a Science of Mythology*, vol. 4 (New York: Harper & Row, 1981), 549.
41 Claude Lévi-Strauss, *Tristes tropiques* (New York: Atheneum, 1968), 198–99.
42 Sottsass, translation of catalog essay in *Architettura naturale*, xxii (see note 3).
43 Fumihiko Maki, "Primary Architecture," in *Emilio Ambasz Inventions*, 45 (see note 1).
44 Roland Barthes, *Mythologies* (New York: Hill and Wang, 1972), 78–79.
45 Ibid., 139, emphasis in original.
46 Ambasz, "I Ask Myself," 27 (see note 2).
47 Ambasz, "Coda: A Pre-Design Condition," 50 (see note 2).
48 Ibid., 51, emphasis added.
49 Ibid., emphasis added.

Mycal Cultural and Athletic Center

Shin-Sanda, Japan, 1990

A cultural and athletic center in the new town of Shin-Sanda benefits not only its employees but also the growing local community. The significant challenge of this project was to accommodate the immense massing requirements of the building's 450,000 square feet, while sensitively acknowledging the serene open landscape beyond. Ambasz succeeded in returning to Shin-Sanda virtually all of the greenery that this enormous building footprint would otherwise have taken away. The Japan Housing Authority highly approved of the overall benefit that such a design solution would bring to the community, and as an incentive and ultimate reward for this inventive solution, they reduced the cost of the land by almost two-thirds.

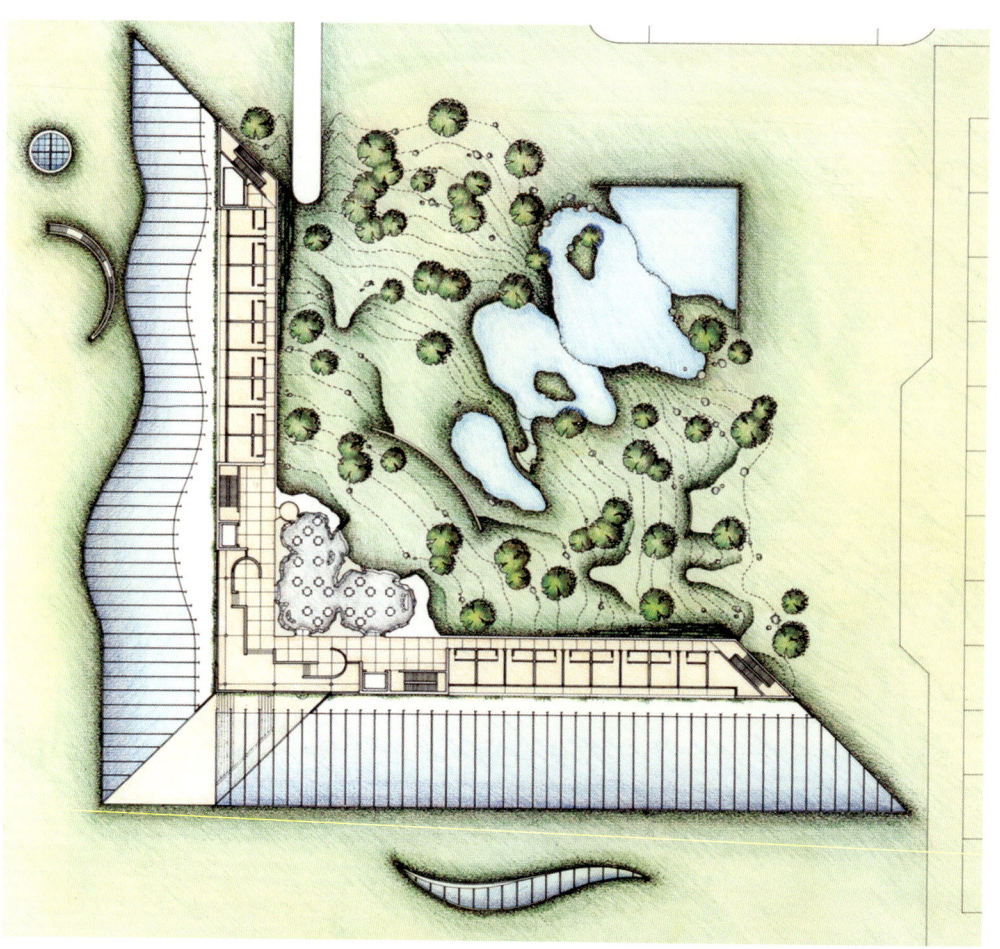

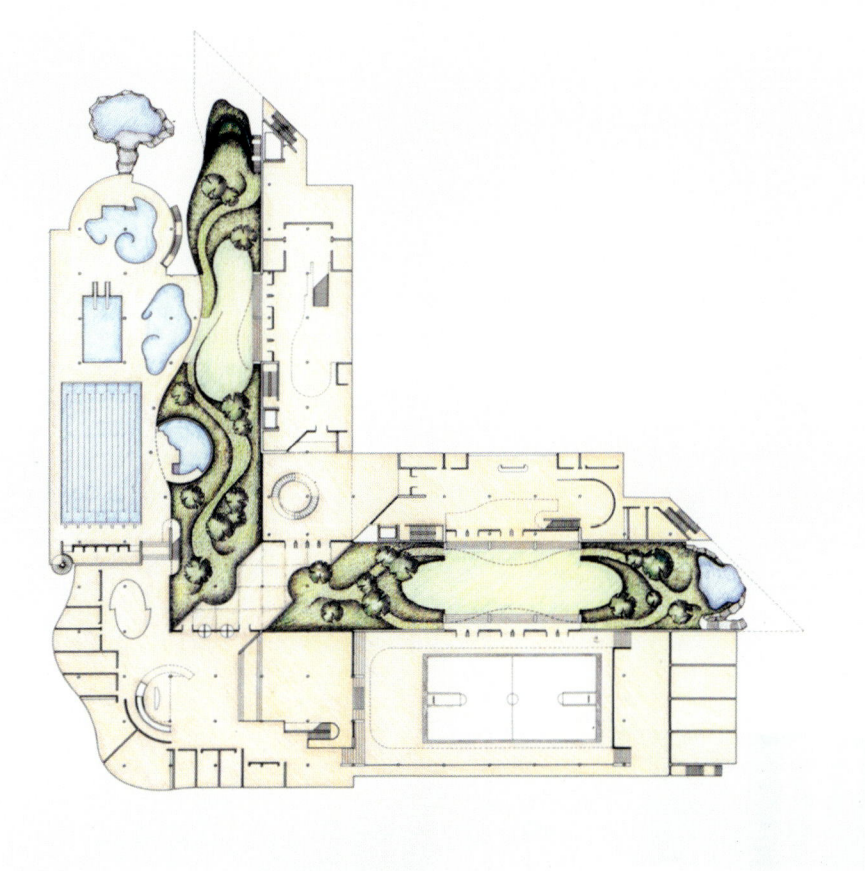

Ambasz refuses to consider technology as a thematic concern, reformulating the greenhouse typology by adding innovative twists to its iconography and even turning its supposed function on its head.

Fulvio Irace

ARGENTINIAN AESOP

Fulvio Irace[1]

Emilio Ambasz's architecture, informed by a fascination for foundation myths' eternal narrative, introduced its eccentric imprint of poetic storytelling into the 1970s architectural landscape. Building on the fabulist aphoristic inventiveness of his "working fables," this singular Argentinian Aesop's first architectural design concepts are actually visual illustrations to accompany a design didactic rooted in an exhortation to rethink the ethical motives underpinning the act of construction, with a view to restoring landscapes disrupted by global transformation and societal changes. As Ambasz puts it, "I opted to be a fabulist rather than an ideologist because fables retain the ring of immutability long after ideologies have wilted. The invention of fables is central to my working methods; it is not just a literary accessory. Sometimes the fable conveys an overarching design concept and the descriptive literary part is purely technical, and sometimes the imagery becomes the illustration. In any case, the subtext of a fable is a ritual and it is to the support of rituals that most of my work is addressed."[2]

In expressing a utopia that bends technology to attain its own goals, "Emile" Ambasz's "narrated" design concept ironically assumes the *conte philosophique's* didactic intonation. Echoing accounts of eighteenth-century travelers, ethnologists, and explorers or the writings of the great Enlightenment moralists, the landscape of an idealized Arcadia—be it an open landscape or constructed urban nature—provides the setting for a human drama; miniature timeless rituals are reproduced in the micronarrative of each design concept, and the main characters and bit players become archetypical recurring figures, as in a fairy tale reinterpreted by Propp.

Set between preindustrial human values and the logic governing serialized production, the ideal inhabitants of Ambasz's fables step into the role of figures seeking to push beyond the constraints imposed by the products and progress of their culture, while realizing that it is precisely the legacy of such phenomena from the past that must be deployed in creating a new culture. The context is formed not by history, but rather by the absolute space described by Martínez Estrada in which each individual "is alone, like an abstract being that will begin anew the story of the species—or conclude it."[3]

As the last historical subjects and progenitors of a new anthropology, humans are thus compelled to draw on myth's indemonstrability or the simplification of fable in order to resolve the ambivalence that the human condition entails. "In choosing to be a fabulist rather than an ideologist," writes Michael Sorkin,[4] "he has grasped something fundamental: that fables preserve their immutable character long after ideologies have faded away. For Ambasz, irony is the most productive critical faculty. It does not attempt to adduce something definitive, only to find a good response."

Fable versus Ideology
Building a bridge between Europe and America, Ambasz's architecture thus quietly debuted in an ambience marked by both technological utopia's last glimmers and the first signs that the discipline was being reduced to merely its decisive archetypes and figures. Nevertheless, Ambasz's choice of metaphor as the frame for his design concepts is in stark contrast to the ideological strategy that was for example crucial to Aldo Rossi or Carlo Aymonino's reformulations of urban theories in Italy, and also informed the progressive consolidation of a school devoted to the cult of typological permanence and its indifferent constancy on the ground. Reaffirming the autonomy of architecture—with its familiar corollaries of independence from function, the establishment of primary elements such as monuments and the value attributed to these in terms of collective memory, etc.—gave unexpected contemporary relevance to the prescriptive knowledge stratified in the literature, treatises, codes, and theories, denying the break with history deployed by modernity to legitimate the twentieth century's tabula rasa tactic. At the same time however, this approach rejects attempts to create a cross-hybridization between architectural codes and their purportedly invalid counterparts in fields like urban planning, sociology, or psychology.

    The young Argentinian was familiar with the practices described above thanks to his contacts with the Italian architectural scene while preparing the famous exhibition *Italy: The New Domestic Landscape* for the Museum of Modern Art in New York in 1972. That was also the year that the "Five Architects" of New York became a group, at the instigation of Arthur Drexler,[5] subsequently affirming a "poetics of nostalgia" which set a sophisticated interpretation of the modern movement's intellectual and formal legacy at the heart of their own experiments with architectural idioms.

    The reading of the Italian situation Ambasz presented focused essentially on the contraposition of the apocalyptic and the integrated in industrial design practice; reacting against a purely formal interpretation, he underscored the emblematic and allusive value of Italian design's participation in social discourse. Although he took on board the notion of a "negative utopia"⁶ adopted by some strands of the radical avant-garde as a tool for analyzing the present, he moderated its scope through the prism of a more balanced relationship with the functional component of elitist professionalism, highlighting in particular the crucial role played by politics and aesthetics as complementary elements in an organic design vision aimed at reestablishing the discipline. This diagnosis was fairly close to conceptions of the "poetics of the pragmatic," his personal take on which soon appeared in statements on his subsequent design concepts. "Thus, design ultimately transcends both object-making and conflict to encompass all the processes whereby man gives meaning and order to his surroundings and his daily patterns of life. [...] Without claiming to solve everything, design can nevertheless move man toward an authentic realization of himself."⁷ The exhibition catalog that accompanied *Italy: The New Domestic Landscape,* grounded in an interweaving of history, criticism, and projects, can thus be seen as Ambasz's first attempt — except for his early research texts on Buenos Aires — to construct an image of the architect as a fabulist, moving between words and things.

Restoring the image to the evocative status it assumes in myth also implies reinstating the theoretical and utopian dimensions of the overarching design concept as a tool for transformation, over and above the constraints of actual built structures. It entails highlighting the idea of invention through the notion of the prototype as an anticipation of a potential future, hinting at a greater openness to new symbolic representational systems, reflecting the contemporary world's contradictions and aspirations.

Despite its attraction to utopia, Ambasz's architecture escapes any compulsion to follow the route of technological visions, a seemingly compulsory way station for the international avant-garde, who have adduced the most varied arguments to underpin their phantasmagoric inflections of such visions, which have spanned the broadest conceivable spectrum, inspired not just by idolatry of the machine, but also by Situationist anarchy or cosmic futurology. Technology however does not assume the kind of redemptive value proclaimed by the avant-garde for Ambasz, who views it merely as an instrument at the service of design imagination: it is an image per se, but rather a structure that enables creation of emotional images. At the Center for Applied Computer Research, it is the configuration of precise canalization and drainage systems that makes the "floating" buildings possible, and underpins the ingenious system that recycles water from the air-conditioning to create the liquid "carpet" that flows slowly and continuously by the steps at the Grand Rapids Art Museum, Michigan. This strategy pays tribute to Amancio Williams with his eccentric, isolated views, his antihistoricist approach to "create and invent exemplary prototypes"[8] such as the Bridge House at Mar del Plata and his deft concealment of his program's "poetic core" under the mantle of the functional paraphernalia of the underlying design concept.

Elucidating the methodology of his inventive practice, Ambasz explains: "An architecture in a dynamic consonance with nature, made by man, in a state of constant becoming involves specialized tasks. The first is empirical: to establish a cartography of the products and production techniques that populate the man-made garden. The second is normative: to develop a program of individual needs and wishes in the context of a far-reaching program of social necessity, to guide utilization of the empirical cartography. The third is synthetic: giving form to new structures that will allow man to reconcile his fears and desires with the limits imposed by the empirical realm and the pressure of the normative sphere. […] The sphere in which the architect operates may change, but the transcendent task remains the same: to give a poetic form to the pragmatic."[9]

Committed to the twenty-four-hour frame that defines everyday life, the Argentinian's "domestic utopia" serves as the metaphor for a transformation more intimately related to anthropological structures than to any architectonic foreshadowing of a renewed form of living space; it thus verges on a poetics of the archaic, whose "frugal vocabulary" refers back, more or less directly, to Luis Barragán's architectural ideas, addressed in one of the first critical essays devoted to the Mexican architect and explored in an invaluable exhibition curated by Ambasz for MoMA in New York.[10]

Between Borges and Barragán

Amancio Williams, Luis Barragán, Borges: even against the heterogeneous backdrop of 1970s New York, the erratic idiosyncrasy of Ambasz's points of reference set him apart, as did the trajectories he followed in conceiving his design approach. Although he had affinities with Peter Eisenman's teaching and was among the founders of the Institute for Architecture and Urban Studies—the American institution most receptive to European neo-avant-garde ideas in this period—his elitist, enigmatic air created a distance between Ambasz and his contemporaries. Among the generation of young Latin Americans who had emigrated to the East Coast, such as Rodolfo Machado and Jorge Silvetti or Diana Agrest and Mario Gandelsonas, he stood out due to his instinctive rejection of critical intellectualism profoundly colored by Structuralist theories on language and possible applications of such insights to the urban context. Ambasz was born in Buenos Aires—the South American capital most open to the influence of international modernism in the interwar era—and studied at Princeton University (New Jersey, USA), where he completed a bachelor's and a master's degree in architecture in just two years. Unlike his American and Argentinian peers, he seemed right from the outset to have little interest in moving into university teaching or the kind of research that seemed to him to entail simply sophisticated but abstract exercises of intellectual rigor. One of his first important works, the 1975 Casa de Retiro Espiritual (House of Spiritual Retreat) in the countryside near Seville, conveys a clear sense of how idiosyncratic his stance was; Ambasz set up an almost didactic comparison with the East Coast elite's design experiments by presenting a solution to a typological theme that had attracted renewed interest from figures such as Richard Meier, Peter Eisenman, John Hejduk, and Michael Graves, who would be key players in later endeavors to reinvigorate architecture. These experiments included, for example, Meier's elegant exercises from the second half of the 1960s engaging with the legacy of European modernism, culminating in the famous 1974 Douglas House in Harbor Spring, Michigan. Meier's ethereal domestic architecture, a refined exegesis of Le Corbusier's theories on breaking architecture down into its component parts, provides a subtle link to the selective reappraisal of European rationalism's hallmarks, a few years before the postmodernist deluge. Carefully reformulating the dialectic between volume and structure, Meier's work aims to recreate the aura of the *esprit machiniste,* as is apparent too in his adoption of white as the sole tonality, subsequently the incontrovertible cipher of his oeuvre. It comes as no surprise that "Whites" was the defining title chosen by this group of 1970s architects—such as Eisenman, Gwathmey Siegel, Seligmann—who shared a great interest in the European rationalist tradition's explorations of architectural language and its confidence "in the capacity of forms and space to mediate between individual and environment, individual and society."[11]

Consequently, as has been noted, this "heroic" season took a turn into the limbo of severe formal reduction, shielding the architect's quest from any anguished search for identity that did not overlap with the exclusive, aseptic focus of linguistic experimentation. This could be dubbed "test-tube" architecture, reinvigorating the tenor of modern tradition beyond the banal consumerism of developers' policy, deftly avoiding any questioning of the troubled nature of the architectonic structure, which remained impeccably sealed in its immaculate white shell.

For almost all these architects, this is generally viewed as just a transitory phase, which they would bluntly reject shortly afterwards or drastically recalibrate in the most extreme form of radical criticism directed at modernity's theoretical postulates. Think for example of Peter Eisenman, an advocate of architecture as Conceptual art, who was deeply engaged in processes of abstraction and self-reflection that aimed to progressively undermine the foundations of the entire theoretical dispositive underpinning the traditional idea of the design concept. As a hermetic experimental architect, the "first" Eisenman identified the single-family home as his almost exclusive experimental field: through his decade-long series of houses, from House I in Princeton (1967–1968) to House XI in Palo Alto (1978), Eisenman constructed a space to reflect on the syntax of his design vision as a solipsistic form of self-representation. Striving to cut rationalist theory's umbilical link between form and function, he analyzed the concepts of centrality and planarity, focusing on relations between individual elements and the structural grid, and disrupting the idea of space and its perception. His idea of the house, which avoids any kind of anthropological twist, is articulated in a set of rules on formation and transformation that placed architecture on a purely intellectual plane, distancing it from the significance of human presence vis-à-vis the form and use of space, as if such considerations were improper or irrelevant. Describing House III in Lakeville, for example: "In this sense, when the owner first enters 'his house' he is an intruder; he must begin to regain possession—to occupy a foreign container. In the process of taking possession the owner begins to destroy, albeit in a positive sense, the initial unity and completeness of the architectural structure."[12]

Real "machines in the garden," Miller House (House III) in Lakeville (1968–1971), like Douglas House (1971–1974) by Meier or Benacerraf House by Graves in Princeton (1969–1970) represent in different ways the return to an idea of art as a "liberation from the eternal essence of the human being" as Tafuri observed[13]: an evocation of suspended tonalities in which silence becomes a synonym for programmatic elision of human sounds.

La Casa de Retiro Espiritual (The House of Spiritual Retreat)

The surreal note that runs through Emilio Ambasz's architecture in those years is immediately apparent in this design set in the countryside outside Seville, split into the dialectic structure of the upper "mask" and the belowground dwelling: steeped in auratic mysticism, it can nonetheless be read as a tacit manifesto advocating elimination of architecture by exhuming its relics. The facade, as a metonymic "surrogate" for architecture, is at once a signal and a ruin: a proclamation that architecture is on the verge of disappearing or a residual trace of its destruction. In the 1972 MoMA exhibition, Gaetano Pesce imagined a habitat that would express the idea of architecture "in the age of great contaminations."[14] The proposal played with the layout for an existential space whose principal physical traits evoked reclusion and isolation: its open, multifaceted symbolism suggested the notion of the house as a ritual locus or a refuge from life, an uneasy domestic landscape or a symbol of eternal life. The indeterminacy of the allusions was however at odds with the clarity of the compositional layout, which drew on the square and rectangle as its fundamental forms. The imposing geometry, based on rotation of quadrilateral forms, produced symmetrically articulated volumes that evolved out of the primitive mark of the cross traced into the ground. While an echo of this can perhaps already be identified in the design for the belowground chapel in the Borrego Springs project—even in the modified iconography of the cross, in this case more evocative of natural religion than of a tombstone—it is in Ambasz's design concept for Seville that reflections on the house as refuge take center stage. A square form dug into the ground specifies the intervention's bounds, and divided diagonally into two triangles that correspond to two sections of the house: one more intimate and on a domestic scale, the patio; the other—a set of steps—is more monumental and tied into the facade's representative order. On the one hand the house thus exists as a volume in space, while on the other hand it is a tangible expression of a space within which life unfolds. This contrast references the distinction between *Baukunst* and *Architektur* that Loos introduced in his famous 1910 essay, which Ambasz taps into on an almost literal note to clarify his conception of the house as the primordial haven, defining the underlying design concept as a pact of reconciliation between humans and nature. Against the backdrop of the recurrent "machine in the garden" phenomenology, the Casa de Retiro Espiritual opts not to foreground analysis of architectural language, remaining indifferent toward mathematical furor that might seek to couch the design concept as a theorem to be demonstrated. Is it just a coincidence that the roots of this design suggest instead the "emotional architecture" of Barragán, a master practitioner of the discipline who assumed an eccentric stance toward the principal trajectories in architecture pursued around the world? In any event, Ambasz deeply admired his work, which he analyzed in detail while researching the MoMA show in New York. For the Mexican architect, the peerless creator of a "stage architecture" who "emphasized living in patios, behind walls,"[15] the house was the locus of his pièce de resistance: his house in Tacubaya (1947), a glorious composition of "voids" behind the explicit modesty of an unremarkable facade, demonstrates how "rudimentary elements" can give rise to an eloquent expression of a powerful "visual drama."[16] Mexican tradition, stripped of any historical imprint, is thus drastically reduced to its spatial underpinnings: a journey back to the immaculate purity of the archetype takes on the cadences of a fresh start, imbuing the new architecture with the "aura of inexorability that classical myths once possessed."[17]

 The two thin walls that meet at an angle to form the Casa de Retiro Espiritual's facade can therefore be read almost as fortifications: minimalistic oversize structures, their function only becomes visible when viewed from the building's lower level. Two steep stairs define a triangle that culminates in the very high mirador or outlook platform, configuring the entire facade as an observation post open to the landscape; its base, in contrast, puts the finishing touch to the patio's square form, emphasizing the centripetal role played by the void within the overall composition. In this sense Ambasz's reading of the Tacubaya House can also be applied to the configuration here—"The garden is enclosed by high walls on three sides; the fourth side is defined by the rear facade of the house"[18]—as can his interpretation of the El Pedregal residential complex, with the garden as the "soul of the house" and the rooms as "retreats meant just for sleeping, the storage of belongings and shelter from hostile weather."[19] A true "program of metaphysical imperatives,"[20] La Casa de Retiro Espiritual's architecture of landscapes dramatizes the metaphysics of the everyday evoked by lines from Borges: "Patio, channel of sky./ The patio is the window/through which God watches souls./The patio is the slope/down which sky flows into the house."[21]

In contrast to the eighteenth-century myth of the primitive hut, architecture is no longer to be assimilated into nature's rationality, but reinstated in its vocation of providing an artificial refuge, drawing on the most elemental gestures of adaptation to the surrounding environment. Ambasz has repeatedly underscored that "In my architecture I am interested in the rituals and ceremonies for the twenty-four hours of the day. I am not interested in rituals and ceremonies for very long voyages, voyages that can take forty or fifty years. [...] And what a tragedy to discover that for the sake of such long-term dreams we have sacrificed our daily lives. No, I am interested only in daily rituals: the ritual of sitting in a courtyard, slightly protected from your neighbor's view and the strong wind, gazing up at the stars; [...]. Dealing with these types of situations attracts me. The ritual is not in the house; I don't make it. And yet the house provides a backdrop."[22]

The Universal Garden
Standing in silent isolation, on terrain that approximates an idealized vision of the Argentinian pampa in the first presentation renderings,[23] the house north of Seville is therefore not a "machine in the garden." Instead of the building being set down within the landscape, it constructs its own landscape in which the sublime and the picturesque are conjoined with a surreal twist that recalls Mary Miss's explorations into the concept of landscape. In her famous 1978 work, *Perimeters/Pavillions/Decoys*,[24] a six-meter-high tower, set on the edge of a field, simulates the frame of a building. At its center, a well, almost five meters deep and delimited by a wooden truss, evokes an ambulatory around an underground courtyard, poetically echoing the Heideggerian metaphor of the "sheltering earth."[25]

As in the American artist's work, the Casa de Retiro Espiritual, beneath the evocative image it creates, conceals the meticulous reality of an artifact that is constructed as a landscape simulation: the empty center of gravity is belowground, the facade rising up like a banner visible from afar. For all the intractable geometrical clarity of its sharp angles, the Casa nonetheless encompasses an irregular organism, emerging in places from the earth berms in the form of sinuous incisions in the turf.

An astronomical observatory, a kind of sundial, offering a vantage point from which to survey the geography of the land, corresponding to a vision of architecture as "one aspect of our quest for cosmological models" and confirming the idea of the design concept as aspiring to reflect the desire "to possess at least an attribute of the universe."[26]

This discontinuous backdrop, the iconic symbol of the house, is so abstract that it looks like an oversize sculpture marking the site: a landmark that conveys a sense of the landscape not as an entity in its own right but as produced by a process of perception grounded in dynamism and change. It only becomes clear that the facade is a simulacrum when you walk around the corner, and it is only as you move closer to the threshold to the patio that glimpses of an enigmatic iconography appear: the undulations and piercings of the ground, the skylights that allow natural light to penetrate deep into the house and the thin curved membranes that contain the stairs. As the asymmetric center of a reinterpretation of the landscape as garden, it therefore conjures up clear allusions to the "happy, universalizing heterotopia" that Foucault describes as "the smallest parcel of the world" but also as its "totality."[27]

When the first images of the Casa de Retiro Espiritual began to circulate, Charles Jencks had already launched his seminal manifesto lauding postmodernist historicism[28]: however, Ambasz's remains at one remove from such contemporary theories, a stance verging on proud isolation, for all his proclaimed attention to the primordial values crystallized in richly evocative recurrent archetypes. The reference points for his intellectual standpoint do not lie within the tradition of debate within his discipline but rather in the American avant-garde's artistic explorations in the 1960s, which Rosalind Krauss celebrated in her famous essay "Sculpture in the Expanded Field," in 1979—the year that the Casa de Retiro Espiritual was completed.

Ambasz's "collaborations"[29] with artists form an important chapter in the evolution of his design methodology, not only because the image is the only doorway that opens access to myth's more profound meanings. His admiration for artists such as Robert Smithson, Michael Heizer, or Richard Serra reflects their shared project of attentive reading and construction of places, rethinking the relationship between nature and the environment, as a creative response to awareness of ecological issues highlighted in increasingly alarming terms, from 1970 on, by publications such as the *State of the World* reports.

Defining the patio as the central element in the dwelling and using cuts in the ground as ventilation slits for the interior, the Seville house is a de facto precursor of the current focus on environmental sustainability thanks to the way in which it envisages drawing on natural resources and integrating traditional typologies and building techniques while avoiding any hint of nostalgia.

Echoing contemporary Minimalist experiments by Donald Judd, Carl Andre, Robert Morris, or Sol LeWitt, the exploration of primary forms does not however look back to history, but combines geometrical rarefaction with the industrial catalog's realism: the Center for Applied Computer Research in Mexico City (1975) uses water from the network of canals beneath the city's suburbs to create an artificial lake, given definition by sculptural elements in the form of large, inclined solar panels. In a literal interpretation of concepts of flexibility and growth, the rafts bearing the offices form landscape compositions that can be varied as a function of demand and the imperatives of expansion.

From this point on, water, earth, air thus become recurring motifs in Ambasz's landscape architecture and initially seem almost to signify nature's revenge on human artifacts, which are reduced to the status of noble "ruins," for example in the design concepts for the Grand Rapids Art Museum (1983) in Grand Rapids, Michigan, and for Union Station (1986) in Kansas City. It is however in the proposal for the Schlumberger Research Laboratories (1982) in Austin, Texas, that the full radical impact of Ambasz's natural vocabulary unfurls, calling into question the practice of dense, concentrated construction by deploying the metaphor of the inhabited garden.

Texan Arcadia

This metaphor already figured in the 1976 design concept for a group of Mexican-American grape growers that Ambasz presented in the "Europe/America" exhibition at the Venice Biennale,[30] explaining his approach in the following terms: "Europe is perpetually in search of utopia, the myth of the end. The recurring myth of America is Arcadia, the eternal beginning. Whereas the traditional vision of Arcadia is that of a humanist garden, the American Arcadia has been transformed into man-made nature, a forest of artificial trees and shadows of the mind."[31]

The citizens of Borrego Springs—to some extent a reflection of the "noble citizen" whom Loos contrasted with the rootless citizen's uncertainty—stage the contrast between eternal human needs and the artificial garden's variable landscape. Rather than being symbols of a Romantic view of organic expression as a means to overcome difference, they can be read as industrious Robinson Crusoes looking for an escape route more in keeping with the new posttechnological era. Similarly, in the Schlumberger complex, management and researchers live and work in a landscape that is reminiscent of Anglo-Saxon informal gardens yet devoid of the picturesque iconography this usually entails. Aware of the need to conclude an efficient ecological pact, and indeed a new social pact, they demonstrate how important it is to establish a relationship to the new technology's advantages and future environmental impact that is less mechanical or directly causal. Employees at the Mexico City research center almost constitute the utopia of an ideal community, perhaps closer in its profound aspirations to Ledoux's Enlightenment symbolism than to Yona Friedman's modern-day fascination with technology.

The idea of the inhabited landscape is superimposed on the notion of the anthropized environment, integrating the rich visual legacy of landscape art's experimentations in a highly original fashion. The Austin technology cluster emphasizes above all that the whole is more significant than the individual interventions, drawing into the light the structure of a landscape that showcases "the strata of the Earth" like a "jumbled museum." In the process, it reinvigorates the insights of Robert Smithson's "Earth Projects" on the unprecedented scale of the final realization of the design concept: "Embedded in the sediment is a text that contains limits and boundaries which evade the rational order, and social structures which confine art."[32]

For the architect, as for the artist, intensive analysis of the language of the rocks discloses a syntax of fissures and ruptures. Whereas Michael Heizer "draws" the colossal furrow of *Double Negative* in the Nevada Desert by displacing and reshaping 240,000 tons of sand and rhyolite, Ambasz cuts irregular meandering "slashes" into the garden's grassy surfaces to bring light and air to the underlying spaces. In a counterpart to Heizer's practice of moving earth, here the soil is raised to shape geometrical berms that integrate the architecture into the soil, cutting energy costs and providing work spaces with generous wide vistas.

The history of a domestic, eternal return thus unfolds against the backdrop of the relationship between humans and nature, drawing inspiration from dreams of a rational foundation to reconcile architectural discourse and built structures. Ambasz writes "I believe that in our pursuit to master Nature-as-found, we have created a second man-made-Nature, intricately related to the given-Nature. We need to redefine architecture as one aspect of our man-made nature, but to do so we need first to redefine the contemporary meaning of Nature."[33]

Green over Gray

"Building new cities by repeating old suburban schemes and, even worse, old mistakes—that is not what I was trying to achieve. What I would like to propose and build is a new 'green' city: a city that, unlike traditional models, is not the realm of the house 'in' the garden but would offer the house 'and' the garden. I have spent the last twenty years of my professional life experimenting with buildings that are able to give as much green as possible back to the community, on the basis of a design strategy that I like to define as 'green over gray.' Year after year, project after project, I have continuously worked on this idea, devising a method that entails first of all creating a 'catalog': a typological set of samples that covers the entire functional range of the various buildings needed for the functioning of a new-style green town."[34]

Ambasz, a highly imaginative compiler of "catalogs of the ineffable," therefore works on the materialization of "a recurrent idyll of place, mapping and remapping his private order of signs. [...]"[35] Akin to the Surrealist technique of collage, the catalog strategy is however essential to the search for the foundational principles of construction: in the apologue of "Manhattan, Capital of the XXth Century,"[36] creating a catalog of domestic places combines memory with designing the future. On the one hand, therefore, we find a taxonomy of fragments of the past that have survived although their context has vanished—Japanese terraces to watch the sunset; Roman baths; patios and courtyards; medieval window seats etc; on the other hand, imagination that looks to the future in identifying places with no historical antecedents that reflect spatial concepts such as flexibility, adaptability, territoriality, privacy. Perhaps this obsession with recombining a clearly identified repertory of signs lies at the origin of Ambasz's highly individual stance, indefatigably proposing a vision of Arcadia that flies in the face of the predominant festishization of technology in his profession, as James Wines has observed.[37]

The extraordinarily coherent objectives pursued in Ambasz's projects compel us to view each instance of his architecture as a

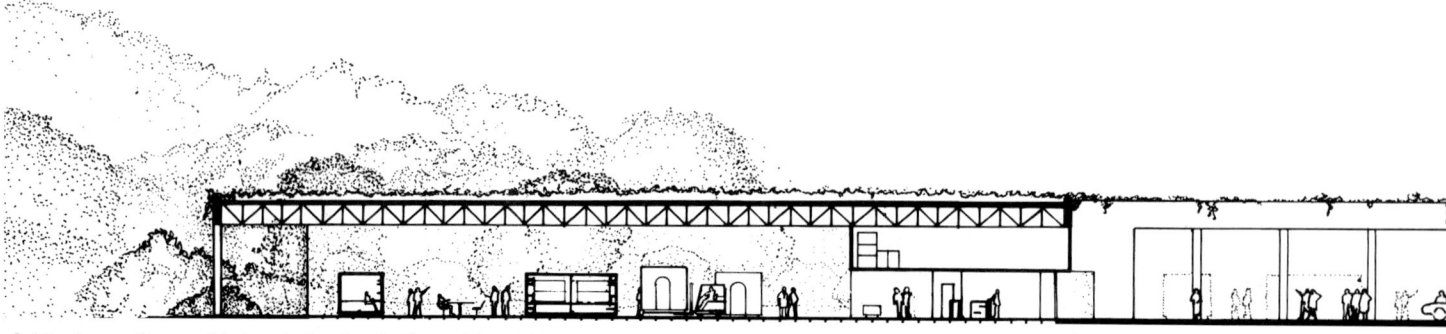

Schlumberger Research Laboratories, Austin, Texas, USA, 1982

further component of his patient research, a figure of a single compositional grammar. In this light, the office complex in the La Venta region, on the outermost periphery of Mexico City, is paradigmatic. Adopting a modus operandi focused on introducing new buildings into a neglected forested area while keeping just as much greenery, the design structures the ground in stepped terraces, with tall trees in the nurseries soaring above the office space. The architecture is not inserted within the landscape because it is actually worked out within the landscape, creating a huge variety of solutions that extend in clusters along a winding access road. Each of these terraced configurations takes on a different form, composing an imaginary topography of rectangular, triangular, trapezoidal, or amoeboid forms, offering a striking synthesis of Ambasz's penchant for conceptualizing geometry as a tool to generate syntheses within a highly complex general morphology. As a result, each of his design concepts unleashes a hint of the primordial gesture of cutting into the earth in a kind of ritual pact of refoundation. The Nuova Concordia Resort Housing Development in Castellaneta, Italy, the Winnisook Lodge in the Catskill Mountains, in New York State, or the Worldbridge Trade and Investment Center in Baltimore, Maryland, are true architectonic geologies that allude to a cosmological vision of the territory, like abandoned grassy cumuli or traces of an ancient civilization excavated from the earth, symbolic manifestations of a ceremony of reconciliation with nature, offering as a pledge to nature architecture that is mitigated: in other words, akin to the notion of the ruin.

A Twentieth-century Paxton

If the architect is not "to find himself a gardener in a man-made desert,"[38] technology must be transformed to be in tune with natural elements. The San Antonio Botanical Gardens, designed in the same year as the Schlumberger Laboratories, define the terms of an environmental pact that works out ecological values in aesthetic values, giving an optimistic and creative connotation to the lamented "death of the landscape."

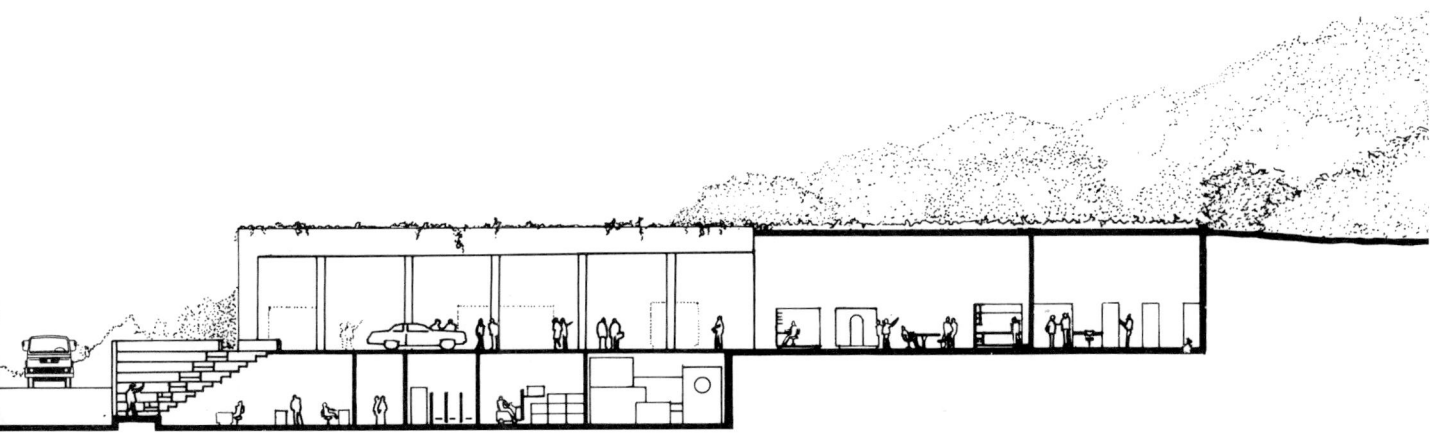

Little more than a decade separates the Lucile Halsell Conservatory in Texas from Nicholas Grimshaw's Eden Project in Cornwall and the Great Glasshouse by Norman Foster in the National Botanic Garden of Wales. Grimshaw's geodesic domes and Foster's sophisticated metallic toroid are extreme reformulations of the robust British technological tradition, updating rather than revolutionizing the nineteenth-century greenhouse, by enclosing the spectacle of nature within lightweight translucid "bubbles" set within the contours of the ground.

Ambasz refuses to consider technology as a thematic concern, reformulating the greenhouse typology by adding innovative twists to its iconography and even turning its supposed function on its head. Greenhouses traditionally protect plants from inclement weather, making the most of every ray of sun; in contrast, in San Antonio, plants must be protected from the sun's excesses, so that the soil becomes the element that actually shelters the plants, while the large curtained glass panes, which can be inclined to various angles as the sun moves, simply serve as a canopy.

Rather than alluding to the future, this techno-ecological vision proposes a view of the entire botanical complex as an archaeological site: an evocative open-air panorama of "ruins" in which the vegetation becomes an integral part of the construction and thus in a sense symbolizes nature's victory over the arrogance of human artifact.

In its clear rejection of biological interpretations of environmental issues, this approach, which has continued to develop in Ambasz's subsequent designs, reveals how very different his most recent work is from science-driven strategies that incorporate ecology by simply applying more refined technological components to stereotypical typologies and conventional built volumes.

Renewed focus on ecological issues has produced scientific protocols and fostered environmental policies, but has not yet been as successful in generating convincing architectural responses. Ecological concerns are now high on the agenda both nationally and internationally, underpinned by increasingly robust technical know-how, and the spread of schools with an environmental focus. Widely associated in public opinion with reasonable expectations of improved quality of life, such concerns simultaneously interact with sophisticated demands stemming from a culture of well-being that celebrates the body as reflecting an almost religious conception of the environment. However, the ramifications of all this have only recently and laboriously begun to filter through into construction typologies and to trigger reformulations of the ways that space is shaped and defined. Ambasz's merit lies in having anticipated and fostered the new alliance between humans and nature as a visionary priority, while at the same time avoiding the pitfalls of nostalgic misinterpretation that would reassert a world of lost forms.

Consistently transposing the principle of environmental remediation into architectonic design, Ambasz has made a hugely original contribution by consolidating a culture of the landscape that moves away from an elitist "art of gardens" restricted to a few exemplary cases and asserts itself as the favored tool in environmental regeneration; many even unhesitatingly identify this as a new paradigm: the design concept as modification. Long before the Dutch tackled the idea of the "hybrid landscape" in the 1990s, Ambasz focused with great precision on the "artificial landscape," not as a simulation of lost innocence but as a working hypothesis to reconstruct a "technological innocence" that can play a key part in reshaping an environment in keeping with the quintessence of the third modern age.

Against the backdrop of highly diverse contemporary engagement with such issues around the world, Ambasz's oeuvre generously anticipates many of the most widespread chararacteristics of what has been dubbed "green architecture." Working in relative isolation from distracted critical opinion, Ambasz has tackled a host of topics in this realm including exploring the potential of underground space, landscape design's importance as a point of reference for architecture, typological inflections of the "green facade," or the "garden" as an informal landscape in urban and territorial regeneration. His engagement with these questions has led James Wines to dub him the "new messiah of environmental architecture."[39]

Fortunately the broader context has now been profoundly transformed, and science and economy underscore the design credibility of Ambasz's "soft manifesto." Although when it was first formulated, this manifesto still seemed to be a humanist hypothesis verging on the idealistic, it can now transpose into blueprints based on a clear design concept the precepts of the entropic paradigm formulated with dramatic clarity by James Rifkin, thus highlighting how far removed Ambasz's oeuvre is from the ephemeral splendor of the hyperconsumerist architecture that left its mark on the second half of the twentieth century. In an architectural landscape increasingly characterized by the perverse fascination of iconicity and an excessive pursuit of theories borrowed from philosophy, Ambasz's architecture has pursued a pathway of eccentric solitude, preferring the originality of a constant, obsessive quest to the uniformization imposed by the critic-driven system to mark architecture as recognizable and affiliated to a particular architectural camp. In the face of the arrogant reception reserved for "square pegs that do not fit into round holes," Ambasz has defied shifting fashions. In the process, he has established a modus operandi that has allowed him, in the space of just a few years, to put seemingly unrealistic notions into practice. Examples include the most emblematic of his built fables, the Seville House, along with the house in Montana, the Lucile

1 All text by Fulvio Irace originally written in Italian have been formerly published in English in Fulvio Irace, *Emilio Ambasz: A Technological Arcadia* (Milan: Skira, 2004).
2 Emilio Ambasz, "I Ask Myself," in *Emilio Ambasz: The Poetics of the Pragmatic: Architecture, Exhibit, Industrial and Graphic Design* (New York: Rizzoli, 1988), 25.
3 Ezequiel Martínez Estrada, "The Lords of Nothingness," in *X-ray of the Pampa*, trans. Alain Swietlicki (Austin: University of Texas Press, 1971), 7, cited in Emilio Ambasz, "Anthology for a Spatial Buenos Aires (1966)," in *Emilio Ambasz: The Poetics of the Pragmatic,* 33 (see note 2).
4 Michael Sorkin, "Arcadia versus Utopia," *Modo*, no. 22 (September 1979): 34.
5 Arthur Drexler, *Five Architects* (New York: Wittenborn & Co., 1972).
6 Emilio Ambasz, "Summary," in idem., ed., *Italy: The New Domestic Landscape: Achievements and Problems of Italian Design* (New York: Museum of Modern Art, 1972), 422.
7 Ibid.
8 Emilio Ambasz, unpublished typescript, 3.
9 Emilio Ambasz, "Una relazione sul mio lavoro," in *Europa/America: Architetture urbane alternative suburban,* ed. Franco Raggi (Venice: La Biennale di Venezia, 1978), 196–197.
10 Emilio Ambasz, *The Architecture of Luis Barragán* (New York: Museum of Modern Art, 1976).
11 Vincent Scully, from *Progressive Architecture*, 55 (July 1974), quoted in Livio Sacchi, *Il disegno dell'architettura americana* (Rome-Bari: Editori Laterza, 1989): 121.
12 Peter Eisenman, "To Adolf Loos & Bertolt Brecht," *Progressive Architecture*, 55 (May 1974): 54–55.
13 Manfredo Tafuri, "Les bijoux indiscrets," in *Five Architects NY,* ed. Camillo Gubitosi and Alberto Izzo (Rome: Officina Edizioni, 1976), 17.
14 Gaetano Pesce, in Ambasz, *Italy: The New Domestic Landscape*, 212–22 (see note 6).
15 Ambasz, *The Architecture of Luis Barragán*, 12 (see note 10).
16 Ibid., 33.
17 Ibid., 12.
18 Ibid., 34.
19 Ibid., 9.
20 Ibid., 108.
21 Ambasz, "Anthology for a Spatial Buenos Aires (1966)," 37 (see note 3). Translator's note: The quotes are from Borges' poem "Un Patio" (1923) and an earlier variant of this poem.
22 Ambasz, "I Ask Myself," 26 (see note 2).
23 Martínez Estrada, "The Lords of Nothingness," quoted in Ambasz, "Anthology for a Spatial Buenos Aires (1966)," 33 (see note 3): "It is the pampa, where man is alone, like an abstract being that will begin anew the story of the species—or conclude it."
24 Christian Zapatka, ed., *Mary Miss: Making Place* (New York: Whitney Library of Design, 1997), 53–55.

Halsell Conservatory in San Antonio, and the cycle of his influential works in Japan. In the past decade the unexpected strand of his fruitful Italian design "tour" (strikingly exemplified in the Ospedale dell'Angelo [Hospital of the Angel] and the Banca degli Occhi [Eye Bank] in Mestre) has been added to the roll call of design concepts that testify to Ambasz's singular vision, whose success is also reflected in the extensive dissemination of his motifs. Although initially his architecture may have seemed simply to foreshadow an intimist utopia that reduces 1960s avant-garde futurological schemes to a poetic dimension, the architect fairly promptly demonstrated that his experiments entail a more mature, professionally aware interpretation of the mission articulated in his overarching design concept, opening up an unexpected window on Adam's eternal dream of returning to his original "house in Paradise."

---

25 Ibid., 10.

26 Emilio Ambasz, "Replies to Michael Sorkin's Questions," in the present volume, 284.

27 "We must not forget that in the Orient the garden, an astonishing creation that is now a thousand years old, had very deep and seemingly superimposed meanings. The traditional garden of the Persians was a sacred space that was supposed to bring together inside its rectangle four parts representing the four parts of the world, with a space still more sacred than the others that were like an umbilicus, the navel of the world at its center (the basin and the water-fountain were there); and all the vegetation of the garden was supposed to come together in the space, in this sort of microcosm. [...] The garden is the smallest parcel of the world and it is also the totality of the world. The garden has been a sort of happy, universalizing heterotopia since the beginnings of antiquity." Michel Foucault, "Different Spaces," text of a lecture presented to the Architectural Studies Circle in 1967, trans. Robert Hurley, in Michel Foucault, *Aesthetics, Method and Epistemology*, vol. 2., ed. James D. Faubion (New York: The New Press, 1998), 175–85, here, 181–82.

28 Charles Jencks, *The Language of Postmodern Architecture* (London: Academy Editions, 1977).

29 See Barbaralee Diamonstein, ed., *Collaboration: Artists and Architects* (New York: Whitney Library of Design, 1981).

30 Raggi, *Europa/America* (see note 9).

31 Emilio Ambasz, "Una relazione sul mio lavoro," in Raggi, *Europa/America*, 106 (see note 9).

32 Robert Smithson, "A Sedimentation of the Mind: Earth Projects," in *Robert Smithson: The Collected Writings*, ed. Jack Flam (Berkeley: University of California Press, 1996), 110.

33 Emilio Ambasz, "Replies to Michael Sorkin's Questions," in the present volume, 286.

34 Emilio Ambasz, "Green Towns," typescript, February 24, 1995, 4–5.

35 Michael Sorkin, "Et in Arcadia Emilio," in *Emilio Ambasz: The Poetics of the Pragmatic*, 17 (see note 2).

36 Emilio Ambasz, "Manhattan, capital del XX secolo," *Casabella*, no. 397 (January 1975): 4.

37 James Wines, ed., *Green Architecture* (Cologne: Taschen, 2000), 69.

38 Ambasz, "Una relazione sul mio lavoro," in Raggi, *Europa/America*, 106 (see note 9).

39 Wines, *Green Architecture*, 69 (see note 37).

I understand architecture as the search for a spiritual abode. On the one hand, I am playing with the pragmatic elements that come from my time, such as technology. On the other hand, I am proposing a certain mode of existence that is an alternative, a new one. My work is a search for giving architectural forms to primal things: being born, being in love, and dying. They have to do with existence on an emotional, passionate, and essential level. Perhaps I use very austere elements to express this quest and, therefore, the gesture may be seen as an austere one also. But by doing it in this way, I believe that it may be far more durable. I am interested in the passionate and the emotional when they assume a timeless guise. EA

VISUAL APHORISMS
Fulvio Irace

I opted to be a fabulist rather than an ideologist because fables retain the ring of immutability long after ideologies have wilted.[1]
Emilio Ambasz

Ambasz has long identified the "working fables" as vital tools in a modus operandi that rejects theory's peremptory tone, preferring the multivalence of metaphor. Theoretical approaches, rooted essentially in rational reflection, are actually totalitarian and absolute; centered on an organizing principle, they construct a process controlled by logical sequences. With its focus on perfect order, theory loses sight of creative disorder in the short span of life, crystallizing the present in the vision of an abstract future, as if it were engaging in preparatory rites for the long journey toward utopia.

As a poetic expression of a condition that is not directly rational, metaphor is expressed in visual thinking and in literary construction, reinstating the image in its role of evoking myths: myths of refoundation that place architecture once again at the heart of the social vocation described by Vitruvius himself as on a par with primitive man's discovery of fire.

As a reworking of the Enlightenment myth of the primitive hut in a postindustrial vein, the grape growers' village in Borrego Springs illustrates how a process comes into being that is animated by the "ritual ethic of growth and renewal."[2] Nature no longer appears as a referent of a rational order destined to be translated into stony solidity, but as a benign model of reconciliation, which mitigates community members' individualistic aspirations by incorporating the ephemeral dimension of the passage of time.

Emilio's island of folly transforms the eighteenth-century penchant for picturesque aesthetics into the narrative frame of a passage describing a private garden and, in constructing a design image, envisages a miniature theater of memory in which the mechanism of memory is analyzed and forgetfulness suspended.

"Emilio's Folly," as well as offering a figurative and allegorical manifesto of its author's idiosyncrasies, is also a catalog of the metaphors that recur in his architecture: water and the earth, the house with a Mediterranean patio, subterranean architecture and the descent toward the depths: to use Sorkin's formulation, it is his own personal "Aleph," "a summary offered with the full certification of the unconscious."[3]

Passing from the canopy at the entrance to the twilight glimmer of the misty cavity offers a didactic description of memory elaborated through cognition's subterranean strata, while simultaneously testifying to the hopes associated with the act of designing, evading paralysis by memory's repetition compulsion.

[1] Emilio Ambasz, "I Ask Myself," in *Emilio Ambasz: The Poetics of the Pragmatic: Architecture, Exhibit, Industrial and Graphic Design* (New York: Rizzoli, 1988), 25.
[2] Michael Sorkin, "Et in Arcadia Ego," in ibid., 22.
[3] Ibid., 17.

Cooperative of Mexican-American Grape Growers

Borrego Springs, California, 1979[1]

Emilio Ambasz

At the heart of this Cooperative of Mexican-American Grape Growers are nine families who, thanks to government grants and loans, have managed to buy some land in a small valley in southern California. The very hot climate is not ideal for grape-growing, but these farmers, advised by viticulture experts from the University of California, draw on a technique used in hot regions in southern Europe. As is customary in southern Italy, the vines are trained up over a wire grid, suspended on three-meter-high columns made of cement and wood, and set at five-meter intervals. The vines grow around each column and then spread out horizontally, so that their leaves create a thick canopy, which protects the grapes from the sun's ravages and allows the shaded ground below to be used for other crops.

The project is divided into four phases. In the first, the nine founding families move to the site in their caravans and live with their families directly under the vineyard's shady canopy: a passageway opened up in two intersecting walls, relics of an abandoned ranch, forms the entrance, while eight square plots, one for each family, will be arranged in a formal configuration, similar to that found in early Hispano-American villages, paying tribute to the families' cultural heritage. Two lines of hedges flank the access route from the gateway to the homes, forming an axis along which the cooperative will hopefully be able to develop and thrive. A small open-air chapel is dug into the ground, with stepped levels descending to the first water table. The chapel's cross emerges from the water, and will shift in tune with its flowing motion. Every Sunday the parishioners will take a spadeful of earth from one of the two heaps at the entrance to the chapel and will shovel it onto a second mound until all the soil forms one heap; at that point a new cycle begins.

Electricity is supplied by a generator connected to a large wheel with wooden blades, set in a pond. A canal dug across the site carries water to the livestock pen, which will be carefully positioned to ensure any animal odors do not carry to the dwellings.

Cooperative of Mexican-American Grape Growers, California, USA, 1976

1 The original text was in English. It appears in Emilio Ambasz, *Architettura & Natura/Design & Artificio; Architecture & Nature/Design & Artifice* (Milan: Elemond-Electa, 1999). This is a slightly amended version approved by Emilio Ambasz.

In the second phase a further sixteen families arrive with their own mobile homes. The organizational principle is different for these dwellings however. They will be set in two new large quadrilaterals, configured to define a triangular plaza for Friday night dances and the market on Saturday where the cooperative's produce will be sold to surrounding villages. Each of these residential quadrilaterals will be divided into nine smaller plots. The central plot in each is a play area for children from the eight families who live around it, while adolescents can spend their free time in the garden on the banks of the pond; this grid, formed by seven-meter-high hedges that create square openings, will be the only example of urban order in the valley's vast expanse. The succession of small private spaces carved out of the hedges' prism-like forms creates isolated nooks and crannies where everyone can enjoy a moment of solitude or meet with friends.

The smallest of the squares facing the triangular plaza is a semi-public space, with an oven and large tables for cooking and sharing midday meals. The tables also function as school benches for classes on Mexican cultural heritage to supplement the children's regular education.

Before the second group of settlers arrives, the cooperative's growing production calls for expanding an underground cellar to be built close to the entrance, providing a cool storage space for the wooden vats. The grapes are collected in a conical silo, a structure with Mexican roots that was traditionally used to store farming produce.

The third phase — imbued with hopes that the divisions into private space for each family will be abandoned, fostering a more community-based lifestyle — is perhaps more a reflection of the architect's wishes than of the settlers' plans.

In the fourth phase we find a metaphor expressing the eternal yearning for all walls to crumble away, allowing humans to live together in peace beneath the shade of the vines, drawing succor from their bountiful grapes.

Cooperative of Mexican-American Grape Growers, California, USA, 1976

Emilio's Folly:
Man Is an Island

1983[1]

Emilio Ambasz

No, I never thought about it in words. It came to me as an image—full-fledged, clear and irreducible, like a vision.

I fancied myself the owner of a wide grazing field, somewhere in the fertile plains of Texas or in the province of Buenos Aires. In the middle of this field was a partly sunken open-air construction. I felt as if this place had always existed. The entrance was marked by a three-column baldachin supporting a lemon tree. From the entrance a triangular earthen plane stepped gently toward the diagonal of a large, square sunken courtyard—half earth, half water. A rocky mass rose in the center of the courtyard resembling a mountain. A barge made of logs floated on the water; it was sheltered by a thatched roof supported by wooden trusses resting on four square, sectioned, wood pillars. Using a long pole, the barge could be sculled onto an opening in the mountain. Once inside this cave I could alight the barge on a cove-like shore illuminated by the zenithal opening. More often, I used the barge to reach an L-shaped cloister where, shaded from the sun or sheltered from the wind, I could sit and read, draw or just think. The cloister was defined on the outside by the water basin and on the inside by a number of undulating planes screening alcove-like spaces. Once I discovered their entrances, I began using them for storage.

Although I thrive on a rather tenuously controlled disorder, I decided I would use these alcoves in an orderly sequence, storing matters on the first alcove until full and then proceeding clockwise to the next one. The first items I stored were my childhood toys, school notebooks, stamp collection and a few items of clothing to which I had become attached. Later I moved gifts I had received while serving in the military, as well as my uniform, out of the house and into the second alcove. I became fond of traversing the water basin once in a while to dress up in my uniform, assuring myself I had not put on too much weight. Not all things stored in these alcoves were there because they had given me pleasure; they were things I could not discard. In time, I developed a technique of using them to support other objects I would put on top of them. I often wondered whether I was going to

1 This version as published in B. J. Archer, Anthony Vidler, Raimund Abraham, et al., *Follies: Architecture for the Late-Twentieth-Century Landscape* (New York: Rizzoli, 1983), 34ff.

run out of space but somehow always found extra room, either by rearranging things or because some objects had shrunk or collapsed due to age or the weight of items accumulated on top.

On the diagonal axis passing the entrance canopy, but directly opposite it, an undulated plane was missing and instead of a storage alcove there was an entrance to a man-height tunnel leading to an open pit filled with fresh mist. I never understood how this cold water mist originated, but it never failed to produce a rainbow.

Emilio's Folly: Man Is an Island, 1983

GREEN FACADE
VERTICAL GARDEN

Fulvio Irace

Through his architecture, Emilio Ambasz does more than simply produce "catalogs of the ineffable": viewed with hindsight, the abacus of his design production reveals the regular rhythm of a combinatory system that adopts an experimental approach, reinterpreting and inflecting recurring typological features or building materials in relation to specific programs.

The "trope" of the metal grid appeared for the first time in the project for the Mexican grape growers (1976) as a metaphor for a sheltering cover, subsequently morphing from this pergola form to become a latticework wall in the Hortus Conclusus maze in the *Parcs et Jardins* exhibition at the Centre Pompidou (1989). It even takes on the role of a "structural" cladding in the New Town Center in Chiba (1989).

Hortus Conclusus, Centre Pompidou, Paris, France, 1989

For a whole host of reasons Ambasz views this city as the prototype of the "green towns" that should become the hallmark of Japan's new urban planning policy, reflecting its highly developed, technologically avant-garde society which is simultaneously sensitive to an age-old tradition of respecting and devoting attention to nature.

Grounded in the principle of melding humans and the environment, the Japanese "green towns" are designed in the light of the great scope for remote working opened up by evolving electronic means of communication. Conceived for at least 10–30,000 inhabitants, these urban centers also have outstanding transport links thanks to the new high-speed rail network. The train station thus becomes the heart of the city, enhancing its configuration as a genuine "town center." In Chiba, for example, the new town center is configured as a multifunctional complex of public and private spaces, with "green architecture" encoding a uniform design process for all the new facades in its urban context.

Ambasz uses an open structural grid as a defining feature within the cityscape, both to mitigate the environmental impact of Chiba's new commercial center and to help minimize the precarious sense of incompleteness that can arise in ambitious long-term construction projects. Space for plants is provided in each module of this three-dimensional metallic network, which Lauren Sedofsky traces back to Sol LeWitt's evocative cubic constructions.[1] The structure thus forms a barrier, shielding the area from nearby road and rail arteries, while also establishing visual continuity between the various parts of the complex; a paradoxical element of permanence in the heart of the evolving new urban center.

New Town Center, Chiba, Japan, 1989

Rephrasing the plant-based grid in its mature form as a fully fledged vertical garden, Ambasz therefore anticipates explorations of mesh facades, notably exemplified in work by Herzog & de Meuron in Napa Valley. His work also foreshadows solutions such as those adopted, for example, by Gaetano Pesce in the Organic Building in Osaka (1994), or by Jean Nouvel in the "plant-based" facades of his design for the French Embassy in Berlin (1997), the "green wall" for the Fiat Belfiore redevelopment in Florence (2002) or the Musée du Quai Branly in Paris (2006) and, last but not least, by Herzog & de Meuron at the Caixa Forum in Madrid.

The Western brise-soleil motif is transformed into the continuous facade of a vertical orangery, planted with various tree species that offer a seasonal cycle of blossoms; a twofold reference, alluding to Western cultural concepts of the well-ordered garden and the Japanese tradition of cultivated nature. The theme is reprised on a colossal scale in the Torii gateway, with two high office towers signaling the entrance to the city as trains pull into the station. The vegetation growing on the facade conceals the underlying curtain wall, evoking Japanese dwellings' covered verandas.

The poetic gesture of transforming the curtain wall into a vertical garden, restoring the natural environment, is also a logical reworking, through the prism of biotechnology, of the focus on energy-saving concerns that has become such a vital part of the contemporary architecture agenda. This is quite deliberately developed into a figurative theme that runs through contemporary commercial buildings, for example in OMA's design for Koningin Julianaplein in The Hague (2002). It also figures in radical green redesigns of this type of architecture, for example in Ambasz's ENI building in Rome's EUR complex, an exemplary testament to the International Style's impact on 1950s Italian design culture. Ambasz's strategy seems on the one hand to embrace the environmental approach that Gabetti e Isola incorporated into the ENI-Snam project in Milan, yet simultaneously proposes an original iconography for the Italian energy giant's headquarters, creating a sense of corporate respect for the natural envi-

Museum of Modern Art (MAMBA), Buenos Aires, Argentina, 1997

ronment and awareness of the need for ecological balance. Wrapped in a double facade that deploys the same vegetation-overlay system that Ambasz experimented with in Chiba, ENI's erstwhile glass palace becomes a metaphor of the new industrial policy of green energy, alluding through its design to the enhanced relationship between the building and its surroundings.

A vertical garden is also carved out of the interior of the erstwhile Tabacalera in Buenos Aires, a nineteenth-century industrial building in brick that forms the starting point for the Museum of Modern Art (MAMBA, 1997). Adopting a method that recalls his highly imaginative museum conversion of the Grand Rapids' Federal Building, Ambasz's design partly hollows out the monumental facade: the brick frontage, adopted as an objet trouvé, becomes a structural grid given rhythm by the columns' positioning. Echoing the subtractions and microdemolitions of Gordon Matta-Clark's X-ray readings of the most private corners of existing buildings, the project removes the window shutters and some walls in the side bays, creating a facade that reveals and simultaneously conceals the buildings that lie behind it. The gaps between the former facade and the new internal walls are filled with a "soft" sequence of trees, giving formal expression to an emerging green facade inscribed in filigree form on the fade-out of its brick counterpart. In the Grand Rapids Art Museum, an inclined plane inserted between the two wings of the building served as a slipway for a silently cascading veil of water. Just as that translucent film conjured up a new semantics for the former public institution's federal architecture, MAMBA's verdant facade plays into a rewriting of the codes for reading the Tabacalera and the anonymous building in reinforced concrete that also forms part of the ensemble. This generates a mise-en-scène of virtual yet still tangibly material architecture, adding a surreal touch to the robust contours of the existing buildings, and setting in motion interactions between artifice and nature, the mineral realm and the world of plants.

1 Lauren Sedkofsky, "Peripheral Vision," *Korean Architects,* no. 131 (July 1995): 12–23.

THE GREEN MOUNTAIN

Fulvio Irace

The first stages of Emilio Ambasz's "journey to the East" date from the second half of the 1980s; the locus of cultural transfers between East and West throughout history, Japan has above all been a lodestone for the dreams of art and modern architecture. From Wright to Taut or Mies van der Rohe, architects have sought, through the mystery of Japanese tradition, to master the tension inherent in modernist endeavors to attain the ideal purity of total "truth."

Is it perhaps just a coincidence that the extraordinary prismatic glass facade in one of Ambasz's first Japanese projects—the Nichii Obihiro Department Store in Hokkaido—is imbued with postindustrial echoes of Bruno Taut's utopian vision, conjured up in Paul Scheerbart's *Glasarchitektur?*

The complex, which has the air of an enormous sparkling quartz, is swathed in a multifaceted glass veil, with tall trees planted in the space between this enclosing element and the structures within. The irregular forms of a greenhouse thus take shape, its contours defined by both the imperatives of building regulations and the precisely calculated interplay of shadows and reflected light.

The trope of the green mountain—recurring so frequently in Ambasz's projects that it in a sense becomes an archetype—is combined here with the image of an inhabited mountain. The iconography created embraces myriad historical illusions, yet is also densely charged with new meanings, as demonstrated by its recent reprisal in two competition designs by Jean Nouvel: the Musée de l'Evolution Humaine in Burgos (2000) and the Guggenheim Temporary Museum of Art in Tokyo (2001).

The Nichii Obihiro Department Store, an optimistic counterpoint to the dense urban fabric in Hokkaido's second largest city, offers a radically innovative twist on the recurrent typology of the shopping mall. This "mountain-mall," as Lauren Sedofsky describes it, rejects both the notion of architecture as container and the idea of a microcity of merchandise, opting instead for a hybrid geological structure in which nature and the man-made pay their design dues to the Japanese art of the garden.

A large hollow center—inspired most immediately by the organic irruption evoked in the Union Station redevelopment project in Kansas City—houses a winter garden that creates an all-pervasive impression of a "naturalistic" landscape rather than simulating the static order of nineteenth-century "palm houses." This symbolic reference to the interior as an organic "cavity" is the first in a family of recurring tropes in Ambasz's large-scale projects, as manifested in variations on this theme in the Worldbridge Trade and Investment Center in Baltimore (1989), the Mycal Cultural and Athletic Center in Shin-Sanda, and the Fukuoka Prefectural International Hall (1990).

Whereas the Baltimore complex strengthens the allegorical deployment of geological morphology, the Mycal Center teases out the idea of constructing a total landscape, with the visible fragments of the functional structure seeming to hold back a topographical explosion within. In Fukuoka, the stair—a symbol of the descent into the depths of the earth in the visual aphorisms—becomes the underpinning for a "natural promenade" that extends the existing park upwards, creating scenographically resonant effects. The stepped profile of this Japanese iteration of an "Arcadian ziggurat,"[1] stratified into shallow terraces gradually ascending from the surrounding park, combines the practical concern of providing natural light to the office spaces with the symbolic mysticism of a "sacred staircase," which runs through the various gardens—for meditation, rest, refuge, etc.—to conclude in the panoramic vista from the upper terrace. The impression created by this vegetation-cloaked building turns it into a metaphor for a gradually dawning awareness of the city's equivocal identity, finally drawn together into a whole in the view over the bay from the belvedere's heights.

The ethical imperative and constructional device informing Ambasz's "green over gray" design strategy lies in giving the community back all the greenery that has vanished due to new construction projects. It is a strategy that draws on a philosophy of subtraction and repositioning typical of landscape art, while also tapping into an idea of architecture that Ettore Sottsass has described as being "like a talismanic instrument of a wager, of a hidden ritual to fascinate some immense natural divinity, [...] a liturgy, performed to obtain forgiveness for the scars we inflict every day on the planet."[2]

However, the shopping and office centers in Shin-Sanda, Fukuoka, and Hokkaido as well as the National Diet Library in Kansai Science City are also elements of a pragmatic utopia articulated by Ambasz—the "green town movement"—that is particularly relevant to the dynamics of urbanization in Japan: "creating new urban settlements which do not alienate the citizen from the vegetable kingdom, but rather create an architecture which is inextricably woven [...] into nature."[3]

1 James Wines, ed., *Green Architecture* (Cologne: Taschen, 2000), 69.
2 Ettore Sottsass, in *Emilio Ambasz: The Poetics of the Pragmatic: Architecture, Exhibit, Industrial and Graphic Design* (New York: Rizzoli, 1988), 10.
3 "Why Not the Green over the Gray?" *Domus,* no. 772 (June 1995): 83.

The formal and functional inflections of these projects epitomize the catalog of urban solutions that the vision of a postindustrial Broadacre City demands: high-speed trains instead of private cars and "mountain-buildings" instead of the isolated skyscrapers imagined by Wright. The National Diet Library, designed as a node in Kansai's new Science City, embodies the living memory of the country's historical culture, making it both a symbolic and functional element in the new urban landscape built on the hills of Keihanna: a "sacred mountain," a ritual tumulus, akin to those where sacred texts were preserved thousands of years ago. As a monumental reinterpretation of the rock/mountain archetype underlined in "Emilio's Folly," the National Diet Library's hillock of earth lies at the origin of the ensemble's configuration, embodying the vital role that collective symbols play in creating well-defined urban identities.

National Diet Library, Kansai, Japan, 1996

THE EARTH AS A GARDEN
Fulvio Irace

The strata of the Earth is a jumbled museum. Embedded in the sediment is a text that contains limits and boundaries which evade the rational order, and social structures which confine art.
Robert Smithson[1]

Scratching, incising, removing layers of soil: the hallmarks of the concept of architecture as a total landscape seem to be the architectural manifestation of the 1960s American avant-garde's endeavors to create "landscape," analyzed notably by Rosalind Krauss in her famous 1978 essay, "Sculpture in the Expanded Field." Collaborations with artists form an important chapter in the bildungsroman of Ambasz's evolving design strategy, in which he does not simply respond to motifs developed by others but figures as an active protagonist.

His friendship with Robert Smithson, and indeed his admiration for the principal representatives of American landscape art, from Sol LeWitt to Michael Heizer and Richard Serra, springs spontaneously from their natural conceptual kinship, which grows out of a belief that "every act of construction is a defiance of nature": "the ideal gesture would be to arrive at a plot of land so immensely fertile and welcoming that, slowly, the land would assume a shape—providing us with an abode. [...] We must build our house on Earth only because we are not welcome on the land."[2]

Environmental art, typically at odds with the construction of spaces, refuses to represent reality by duplicating it. Instead it chooses to recreate reality anew through tangible interventions; defining places, making a mark on the earth, producing installations that seem not to rest on the ground but somehow to emanate from the very soil. This is architecture generated by digging into the ground and displacing the earth, intersecting with preexisting natural elements. In a radical redefinition of the landscape's traditional significance as a backdrop to be contemplated, it becomes the vanishing point in an undertaking that imagines the setting in terms of total theater. Actively including the viewer makes it vital to include perception in the equation, conferring meaning to the way that each specific site is individuated. Introducing the spectator's viewpoint implies reflecting on the landscape's temporal reality as a locus constructed through interactions between human and nature. That makes it impossible to conceptualize nature as somehow separate or existing prior to cultural anthropology. On the other hand, it opens up scope for a narrative of the landscape as the end result of a working process that bears all the typical traits of a foundational rite, as for example in the Borrego Springs' agricultural cooperative.

Like Michael Heizer or Robert Smithson's earthworks, Emilio Ambasz's environmental architecture presupposes drawing up an imaginary topographic study that is informed by the preeminence of the land as the matrix of the future landscape. Sounding out its contours signifies establishing a symbolic—one might even say ontological or ancestral—correlation between geology and psychology, articulated in didactic, apologist tones in the working fables and on a pragmatic note, focusing on environmental remediation in the large-scale projects. The allegorical dimension of these places is perfectly legible only in a bird's-eye view. The layout of the Nuova Concordia residential complex in Castellaneta, for example, sheds the bucolic air associated with vacation villages when viewed from above, revealing instead a complex, almost hermetic configuration of signs, inscribed in the ground like traces of an archaeological dig. No readily recognizable icons can be identified here, bringing us once again to Ambasz's abstract and minimalist architectural sensibility, which draws on asymmetry and geometry to transform the temptation of naturalism into subtly surreal imagination.

Ambasz elucidates his stance in the following terms: "I believe that in our pursuit to master Nature-as-found, we have created a

Nuova Concordia residential complex, Castellaneta, Italy, 1994

[1] Robert Smithson, "A Sedimentation of the Mind: Earth Projects," in *Robert Smithson: The Collected Writings*, ed. Jack Flam (Berkeley: University of California Press, 1996), 110.
[2] Emilio Ambasz, "Replies to Michael Sorkin's Questions," in the present volume, 284.
[3] Ibid., 286.

second man-made-Nature, intricately related to the given-Nature. We need to redefine architecture as one aspect of our man-made nature, but to do so we need first to redefine the contemporary meaning of Nature."[3]

In the regeneration proposal for La Venta, on the outskirts of Mexico City, reforestation of a badly neglected woodland area serves as a springboard to create an eccentric spectrum of geometric forms amid the vegetation. The new plant nurseries establish a hybrid typology of terraced levels that accommodate office space within and open-air sections to grow tall trees on the exterior. Inhabited as it is by mysterious presences—a rectangle, a triangle, a semicircle etc.—the La Venta forest is a nonmimetic landscape, almost the architectonic equivalent of Foucault's idea of the construction of a space, making a mark on the landscape by earth-based interventions or vegetation. The Winnisook Lodge hotel and leisure complex in Catskill Park, on a wooded ridge of Belleayre Mountain, is a further project in the same conceptual category. Winnisook Lodge is structured around precise, curvilinear forms that are reminiscent of the remains of ancient cities swathed in luxuriant vegetation, seeming almost to be a three-dimensional extension of the adjacent golf course. A virtual presence is therefore correlated to the architecture's "absence": the hint of an analogy that alludes to nature vanquishing the ruins of the built realm.

The Vathorst shopping center outside Amersfoort sets up formal echoes with the idea of a prehistoric site revealed by archaeological excavations; extending in a deep fault line cut into the ground, it seems a fitting tribute to the Dutch environmental ethos, in a country whose entire history is cast as an age-old tale of the antagonism between the sea and the land.

The Winnisook Lodge hotel, Catskill Mountains, New York, USA, 2000

The Vathorst shopping center, Amersfoort, The Netherlands, 1999

BUILDING IN THE GARDEN
Fulvio Irace

Considering the garden as Emilio Ambasz's "Aleph" and "the biosphere of his greatness of place,"[1] as Michael Sorkin has written, the Argentinian architect can easily be seen as a new twentieth-century Paxton. However, whereas the latter, availing himself of the new technological possibilities of "greenhouse-style" construction in his Crystal Palace design, epitomized nineteenth-century pretensions of rational control of the world, the Argentinian architect seems to accord the greenhouse the role of a poetic iconography for his idea of artificial nature.

The greenhouse, a transparent display case for an artificial, climate-controlled environment, is a translucent refuge that functions as a counterweight to the grotto's penumbrous gloom and the soil's compact opacity. In contrast to the magnificent nineteenth-century blossoming of glass-based structures, the greenhouse therefore cannot be viewed simply as a stand-alone structure, for it functions instead as the principal element in a semiotic system that also encompasses the soil in order to establish the effect of a total landscape—as for example in the Edmond de Rothschild Memorial Museum in Ramat Hanadiv.

Ettore Sottsass has noted that in the architectural work of Ambasz "there are almost never objects plainly resting on earth, as is usually the case in more conventional architecture where buildings are just a statement, and that is all. Emilio's architectural creations are a bit outside the earth and a bit inside it. They are like stone slabs emerging from the earth, or fissures cracking the earth open, rather than attempts at controlling the universe by means of logic or agreed-upon signs. His is an architecture seeking, almost always, to represent the internal and eternal movement of an all-encompassing planetary geology while at the same time respectfully reflecting local pulses, explosions, contractions, tempests, and deeply welled mysteries."[2]

Within this complex symbology, the Lucile Halsell Conservatory in San Antonio (1982) remains the principal text and the matrix from which a design family with multiple and sometimes paradoxical applications is generated. In the hot, dry climate of southern Texas, for example, greenhouses protect plants from the sun rather than capturing its light and warmth. In this context the earth is the element that really contains and protects the vegetation, with the variable prisms of curtained glass functioning simply as a covering. The configuration of the glazing varies as a function of the sun's angle of incidence, an expressive catalog of geometry, with the panes projecting upwards like frozen fragments of an underground explosion.

As a result, the project imbues the landscape with the touching and vaguely romantic air of a scenic garden: in a gentle undulating meadow dotted with fragile ruins, the vegetation becomes an integral part of the construction and thus indissociable from it; a collective transposition of the obsessions articulated in autobiographical terms in "Emilio's Folly." The vaguely anthropomorphic layout of the Lucile Halsell Conservatory is centered on its open underbelly, a patio—the eternal "channel of sky" lauded by Borges[3]—that acts as a potent lodestone; with its arcades holding in check the greenhouses' air of controlled explosion, it exerts a powerful magnetic pull on the verdant islands of the individual "places."

Ambasz explains "Architecture is, for me, one aspect of our quest for cosmological models. [...] Every one of my projects seeks to possess at least an attribute of the universe."[4]

In his conception of the San Antonio Botanical Gardens as a ceremonial itinerary unfurling above and below the horizon line,

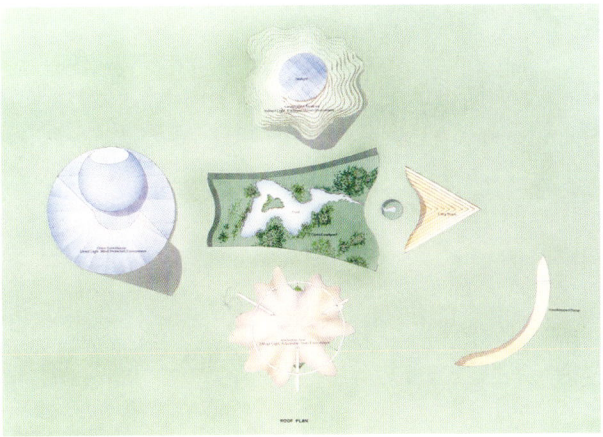

Baron Edmond de Rothschild Memorial Museum, Ramat Hanadiv, Israel, 1993

1 Michael Sorkin, "Et in Arcadia Ego," in *Emilio Ambasz: The Poetics of the Pragmatic: Architecture, Exhibit, Industrial and Graphic Design* (New York: Rizzoli, 1988), 23.
2 Ettore Sottsass, in *Emilio Ambasz: The Poetics of the Pragmatic*, 9 (see note 1).
3 Jorge Luis Borges, "Un Patio," 1923.
4 Emilio Ambasz, "Replies to Michael Sorkin's Questions," in the present volume, 284.

Ambasz demonstrates a grasp of the greenhouse's surreal significance as an expression of the nineteenth-century yearning for the miraculous. This entails mastering the passage of time through technology, optimizing the power of the climate, and bringing together in this virtual Arcadia everything that appears scattered and fragmented in the real world, not to mention seizing pleasure: in a nutshell, eliminating nature's perils and innate harshness.

Perhaps this is why the model of the greenhouse can be applied to such apparently heterogeneous contexts, as Joseph Paxton intuited when he transposed it from the realm of horticulture to create gardens of goods and consumer desires. Beneath the transparent vaults of the Sirmione Thermal Gardens, for example, palms and other tall trees flourish alongside indoor pools, melding physical well-being with contemplation of nature. In the seemingly pristine natural landscape set on a peninsula by Lake Garda, a sense of a charming garden is evoked, yet the most refined wonders of sensory culture are concealed beneath the turf, blending the thermal springs' natural benefits with the artifice of human know-how.

Thermal Gardens, Sirmione, Italy, 1996

URBAN GARDENS

Fulvio Irace

The garden is the smallest parcel of the world and it is also the totality of the world. The garden has been a sort of happy, universalizing heterotopia since the beginnings of antiquity.
Michel Foucault[1]

Set in opposition to the nonsites of utopias, the countersites of the heterotopias described by Foucault mark out the reality of a space in which all the other places of real life "are simultaneously represented, contested and inverted."[2]

Introducing a garden into the historical city's stone-defined order, as in Ambasz's proposal for Plaza Mayor in Salamanca, expresses a contradiction that can be resolved only by a poetic gesture of reconciliation. Breaking down the rectangle's perfect geometry by introducing a series of converging steps into the protected, tree-lined space of a sunken courtyard implies creating an "other" place that interacts with the stony hierarchy of Churriguera's facades, opening a surreal window onto the blithe realm of nature. As a consequence

Eschenheim Tower, Frankfurt, Germany, 1985

of the project's metaphorical value, it can be transposed to very different contexts, equally informed however by the artificiality of the urban landscape. Think for example of the Houston Center Plaza: an open piazza in the perfect grid that constitutes the matrix of the city, simultaneously emphasizing and questioning its historical hallmarks of regular, dense blocks. In both projects the functions of urban life are banished to extensive spaces carved out below ground level, while water features and the striking use of topiary clearly proclaim the role of the senses in perceiving the city.

Ambasz however refuses to define himself as the romantic heir of an Arcadian, antiurban vision, linking his work instead to the testaments of history: "To this day, you can go up to the top of the towers of a medieval city like Bologna, and discover that behind the facades defining treeless streets exist immense gardens which occupy almost 25 percent of the city area. Those were once vegetable gardens, and places where cows grazed. Those grounds were of utmost importance to survive a siege. […] I strive for an urban future where you can open your door and walk out directly into a garden, regardless of how high your apartment may be."[3]

Lauren Sedofsky writes that in fact "in Ambasz's case the urban/nonurban distinction seems null and void. His production involves a prototype developed, not in the country, but in total isolation, a pure laboratory product."[4]

In a sense akin to Robert Smithson's "land reclamation," Ambasz's aspiration to create an elegiac, happy Arcadia does not seek to deny the reality of what exists but instead asserts his focus on a palpable notion of the earth. His more recent work addressing transformation of strategic sectors of the city appears to focus especially clearly on this aspect. That is particularly apparent in the project for the Eschenheim Tower in Frankfurt, which aims to resolve the problem of traffic along the nineteenth-century ring road. It develops a solution that covers over the road infrastructure by introducing variations into the street levels, and incorporates an artificial hill that functions as a pedestrian bridge, linking the existing park and gardens into a verdant organic network. A strategy thus emerges that advocates fluid solutions to redesign circulation in tune with the logic of green urban planning, overcoming the separation of isolated buildings. The rationalist notion of the grid assumes a rational, vital meaning when applied to a range of artificial ground levels, forming the backbone of a renewed link that draws extensively on biotechnology and natural resources. The proposal for a residential and commercial development along Catharijnebaan in Utrecht is entirely in keeping with this approach; it is couched as a fitting tribute to the country most actively involved over the last few decades in developing an innovative conception of the landscape as a man-made technological Arcadia.

---

1 Michel Foucault, "Different Spaces," text of a lecture presented to the Architectural Studies Circle in 1967, trans. Robert Hurley, in Michel Foucault, *Aesthetics, Method and Epistemology*, vol. 2., ed. James D. Faubion (New York: The New Press, 1998), 175–85, here p. 181–82.
2 Ibid., 178.
3 Emilio Ambasz, "Replies to Michael Sorkin's Questions," in the present volume, 285.
4 Lauren Sedofsky, cited in *Analyzing Ambasz*, ed. Michael Sorkin (New York: Monacelli Press, 2004), 60.

DOMESTIC GARDENS

Fulvio Irace

You always have the sense that behind the walls of these projects are absent presences or present absences. The notion of that which is in front of you and what happens behind the wall has always appealed to me. There is a certain anima or spirit behind the wall.
Emilio Ambasz[1]

Picking up on an age-old tradition, the topic of the primordial dwelling, the dream of "Adam's house" and an investigation of phenomena that preceded the emergence of built forms reemerge and play a crucial role in Ambasz's explorations, reflecting his metaphysical concern with returning to the origins.

Radically pared down to just a handful of elements that reflect the drama of architecture celebrated in its liturgy, the Casa de Retiro Espiritual in the environs of Seville, in Spain, conveys the absolute tenor of a manifesto. The house as a refuge forms a poetic response to the intimacy rooted in the space, and, as Ambasz underlines "is not an answer to the pragmatic needs of man (that is, the task of building), but a response to his passion, his imagination."[2]

From a distance two walls that meet in a right angle announce the house's invisible presence; two steep stairways come together on high in the meditative belvedere; two streams of water descend along channels cut into the walls, flowing together in a semicircular basin set at the center of the patio.

The austere, timeless language purifies emotion in the surreal immobility of waiting: the architecture is reduced to a pure "simulacrum" and the earth emerges as the key player in the patio's serene emptiness. Rejecting static representations of space, Ambasz conceptualizes the house as a journey deep into the notion of dwelling, delving into the mystery of what is behind the wall. As Durand reminds us in *Les structures anthropologiques de l'imaginaire* (1960), "the axis of descent is an intimate, fragile, delicate one [...] we remount (time) and rediscover prenatal calm."

The drama of the descent narrative is tempered in the triangular patio, which sets up something akin to a filter between the exterior, reduced to a mask and a regained sense of living space. It is the notion

[1] Quoted in *Analysing Ambasz*, ed. Michael Sorkin (New York: Monacelli Press, 2004), 24.
[2] Emilio Ambasz in *Progressive Architecture* 58, nos. 4–6 (1977): 233.

of the Eternal Return; of the Rituals of the Beginning. There is nothing monumental in the house's interior, which reveals an irregular and organic space structured by surprising recesses and unexpected projections that delimit the areas where the everyday rites of domestic life are celebrated. Diffuse light floods from the serpentine skylights, which carry a sense of the changing seasons into the interior, while curving walls advance and retreat like membranes, opening up to reveal concealed receptacles.

The landscape seems to appear in its pure state in the Manoir d'Angoussart design for the Lambert family in Bierges, Belgium. It is like a still life of geometric forms composed of the earth, conjuring up an image of a field that offers shifting perspectives. An ivy-clad facade welcomes the visitor like the ultimate destination in a trajectory of mute volumes and solid embankments. Set within a gorge, the house is built entirely within the earth, with grass, plants, and raised landscaping forming its introductory salvo and ornamentation, fusing the house and garden into a single entity.

The Montana house is again primarily embedded deep in the landscape, with the pared-down framework of the detached facade peeping somewhat hesitantly out of the grassy surroundings. Designed for a sophisticated art collector, the bucolic "hut" with six rustic wooden columns punctuating its facade is the first in a series of three follies to accommodate the family, the on-site guardhouse and an art gallery.

Private estate, Montana, USA, 1991

Private estate, Montana, USA, 1991

GARDENS OF MEMORY

Fulvio Irace

"An inexhaustible inventor of metaphors,"[1] Emilio Ambasz has always stated that he believes in architecture as an "act of mythmaking imagination": "It is not hunger, but love and fear—and sometimes wonder—which make us create."[2]

With its focus on representing fundamental principles, reflections on memory are set center stage in his architecture: an ancestral memory that touches on history to tap directly into the eternal enigmas of birth and death.

Pro Memoria Garden is the title and the programmatic thrust defining Ambasz's project to create a living memorial to the horror, torments, and destruction unleashed by war. Every child born in the small town to the south of Hanover where it is located is welcomed into the world with a symbolic gift: a patch of land and a marble slab with his or her name inscribed upon it. Each diminutive garden is reassigned to a newborn when its previous owner dies; a new name then joins the previous name or names on the marble slab. As a living reflection of the town's history, the garden will trace out the destinies

Barbie Knoll, Pasadena, California, USA, 1995

1 A. Mendini, in *Emilio Ambasz: The Poetics of the Pragmatic: Architecture, Exhibit, Industrial and Graphic Design* (New York: Rizzoli, 1988), 15.
2 Emilio Ambasz, "Replies to Michael Sorkin's Questions," in the present volume, 284.
3 Ettore Sottsass, in *Emilio Ambasz: The Poetics of the Pragmatic,* 9 (see note 1).

of the local populace while also testifying to the yearning for reconciliation shared by all human beings, transcending ideological and political divisions. The infinite garden's geometrical layout is guided by the organic pattern of a form that is neither foreseeable nor fixed: reflecting all that exists "as an ever-changing process" to cite Ettore Sottsass, Ambasz's work articulates a quest to find "a constant state of fluidity" to convey dreams, aspirations and fears through the prism of quotidian rituals and ceremonies.[3]

There is also a hilltop of memories in the museum-park dedicated to the ultimate icon of eternal dreams of beauty and youth: the Barbie doll. Reinterpreted as a postmodern version of the classical ideal of femininity—a Californian *kore*—Barbie plays the lead role in this informal Temple to Athena set in the hills of Pasadena: a kind of folk Sanssouci dedicated to the myths and history around this famous doll, showcasing the milestones and turning points that punctuate Barbie's fluctuating fortunes.

After entering an open-air courtyard, visitors arrive in the museum's irregularly shaped hypostyle, where each column corresponds to a version of the doll that symbolizes a particular era and fashion in Barbie's long biography. At the end of the visit, the museum's closed space opens up to the exterior. Here visitors encounter a slumbering vision of Sleeping Beauty embedded in a soft, luxuriant carpet of flowers and foliage: a sensual metaphor of eternal life, the American Beauty of infant dreams emerges like the evanescent fata morgana of an eternal lust for beauty.

Barbie Knoll, Pasadena, California, USA, 1995

GARDENS OF HEALTH

Fulvio Irace

Medical and surgical progress have brought about radical changes in life expectancy in the twentieth century, as well as dramatically improving treatment, preventive medicine, and health care, which have attained standards once conceivable only in utopian dreams. In terms of hygiene and distribution of space in the domestic realm, these developments have primarily been articulated in architecture through residential blueprints grounded in the idea that physical health is linked to healthy buildings. However, these improvements in housing design are not mirrored in a similarly tangible transformation of health-care architecture. Apart from a handful of exceptional projects—such as the Paimio Sanatorium by Alvar Aalto or Le Corbusier's unbuilt Venice hospital project—architectural designs have analyzed in minute detail how to ensure buildings in this sector deliver efficient health-care, yet have largely neglected patients' dignity and their right to receive care in the broader sense of the term.

However, rising life expectancy and the associated goal of shorter hospital stays make it imperative to rethink the quality of the surroundings in health-care settings. At the same time, new social standards have spurred endeavors to strike a different balance spatially, incorporating a new emphasis on aspects like light, color, gardens and vegetation in general, and on incorporating areas for patients to enjoy some solitude or rest. Over the last two decades, experimental projects such as REHAB in Basel or the Meyer Children's Hospital in Careggi, Florence, have therefore paved the way for a more humanist view of health-care architecture, introducing parameters usually reserved to private architecture or to leisure/wellness contexts.

The Ospedale dell'Angelo in Venice-Mestre, a rehabilitation facility with around 660 beds, is the first large-scale project to apply these criteria. It reflects an ambitious attempt to bring together a number of exceptional experimental features, consolidating them in a practical form that can serve as a model for numerous other contexts.

The result is surprising, yet the underlying idea is essentially quite simple. Drawing on the principle of the greenhouse and all-round landscape design in developing this major scheme, Ambasz has conceived an extensive district, which also encompasses the innovative Banca degli Occhi (Eye Bank) alongside the hospital. The ensemble becomes an enormous garden of health, an evocative expanse of green in which the buildings rise up like monuments engaged in dialogue. In many respects, the complex's layout echoes the Lucile Halsell Conservatory in San Antonio, Texas, a true masterpiece of the Argentinian virtuoso's green architecture.

The hospital, the defining element in a huge urban park generously planted with trees, functions as a green barrier to urban development, introducing an entr'acte into a dense, built-up area on the city's outskirts. The building, shaped like an artificial hill, is terraced on the side that faces the park, opening on the opposite side onto an atrium-greenhouse enclosed within a vast expanse of glass. By filtering out noise from the railway tracks, this feature also creates a more comfortable environment for patients in rooms facing southwest. The sloping structure creates a visual link to the two unfolded wings of the Banca degli Occhi, creating rhythmic correspondences that suggest the entire complex constitutes a single extensive garden; a landscape-garden characterized by monoliths that are entirely out of the ordinary, evoking a remarkable sense of surprising presences. Two closed eyelid-shapes define the entrance to the Banca degli Occhi, forming a spectacular gateway to the mysteries of scientific research that reaches out with its forthright symbolism to embrace patients' fears and hopes. Elsewhere in the building the metaphor of the eye is continued, with a 450-seat auditorium symbolizing the pupil, while the Biblical symbolism of "fiat lux"—evoked by Michelangelo in his renowned *Creation of Adam* in the Sistine Chapel—is transposed into an abstract form in the courtyard and the two entrance wings.

Ospedale dell'Angelo

Venice-Mestre, Italy, 2008

This 660-bed hospital offers all general hospital services, plus, in the future, a Proton Beam Therapy and Treatment Center. It is conceived architecturally as an aid to the healing process. Its grand entrance hall — a glassed space more than 660 feet (ca. 200 meters) in length, 85 feet (26 meters) in depth, and 90 feet (27.4 meters) in height — is a veritable winter garden, with trees, flowers, and aromatic plants welcoming both patients and visitors. It is the first fully "green" hospital ever built. Patients approaching the hospital, be it by car, bus, or via the newly dedicated train station serving it, pass a large extension of green exterior, visible from every patient's room, to enter the grand winter garden serving as the reception hall. The reverse ziggurat section of the building ensures that half of its patients have a direct view of the winter garden, while the other half have a personal view to the plants growing outside their windows in deep earth and plant-covered terraced containers. The vision guiding this design was that the building should help allay the fears of incoming patients and also contribute to their recovery in their convalescence. Accordingly, patients can perambulate on a series of dedicated high, terraced platforms overlooking the winter garden. For those patients not able to move far from their rooms, every floor offers lounges with views of the winter garden. To avoid the visual intrusion of a large mass of functional service buildings, all surrounding structures — the administration center, the large parking garage, the mortuary, and the adjacent chapel, as well as all laboratories and operating rooms — have been bermed up on three sides and covered with planting, as are their roofs.

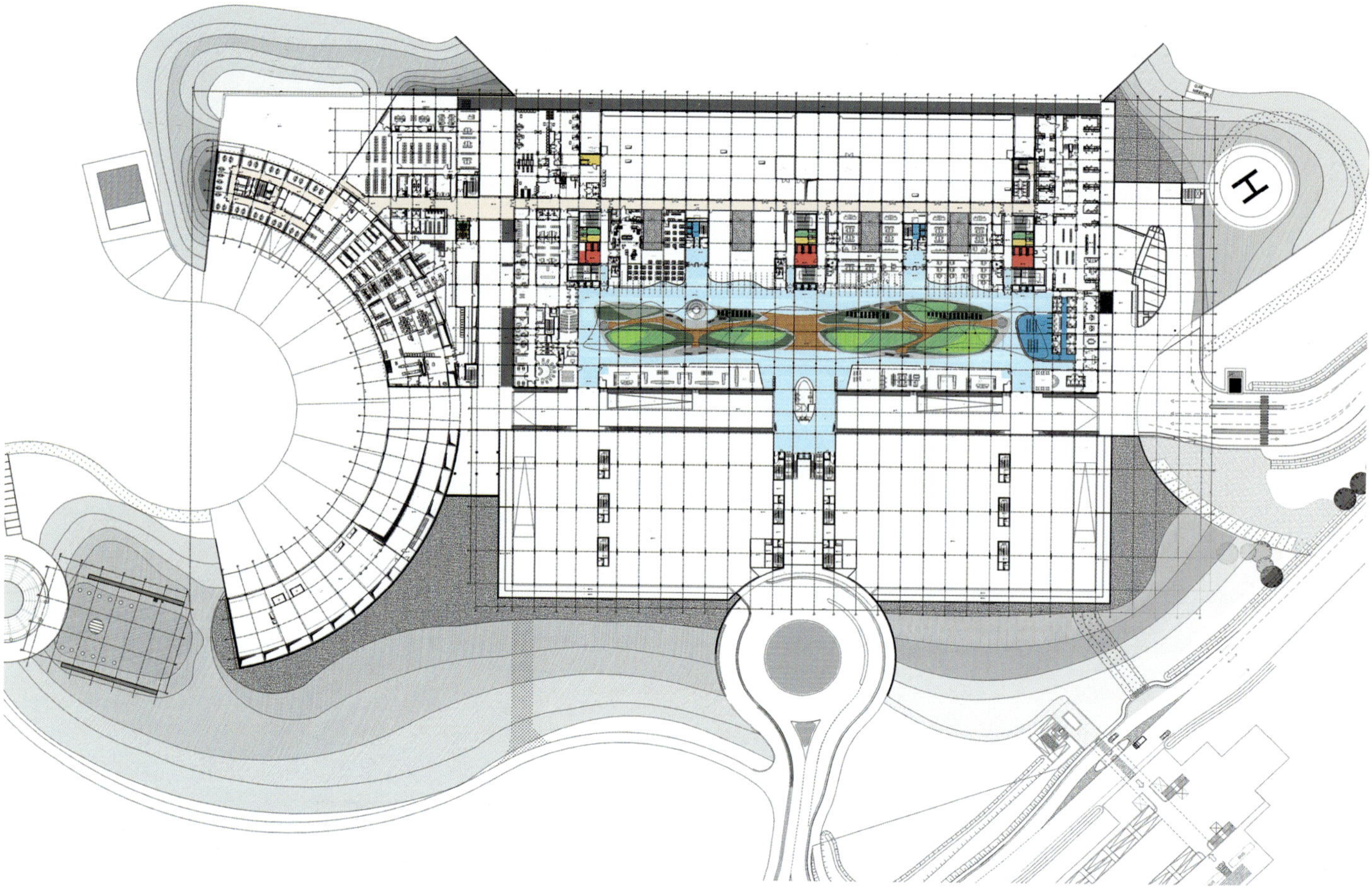

1 Atrium
2 Vistitors lobby
3 Patient rooms
4 Doctor's office
5 Health-care staff rooms
6 Operating rooms
7 Staff parking

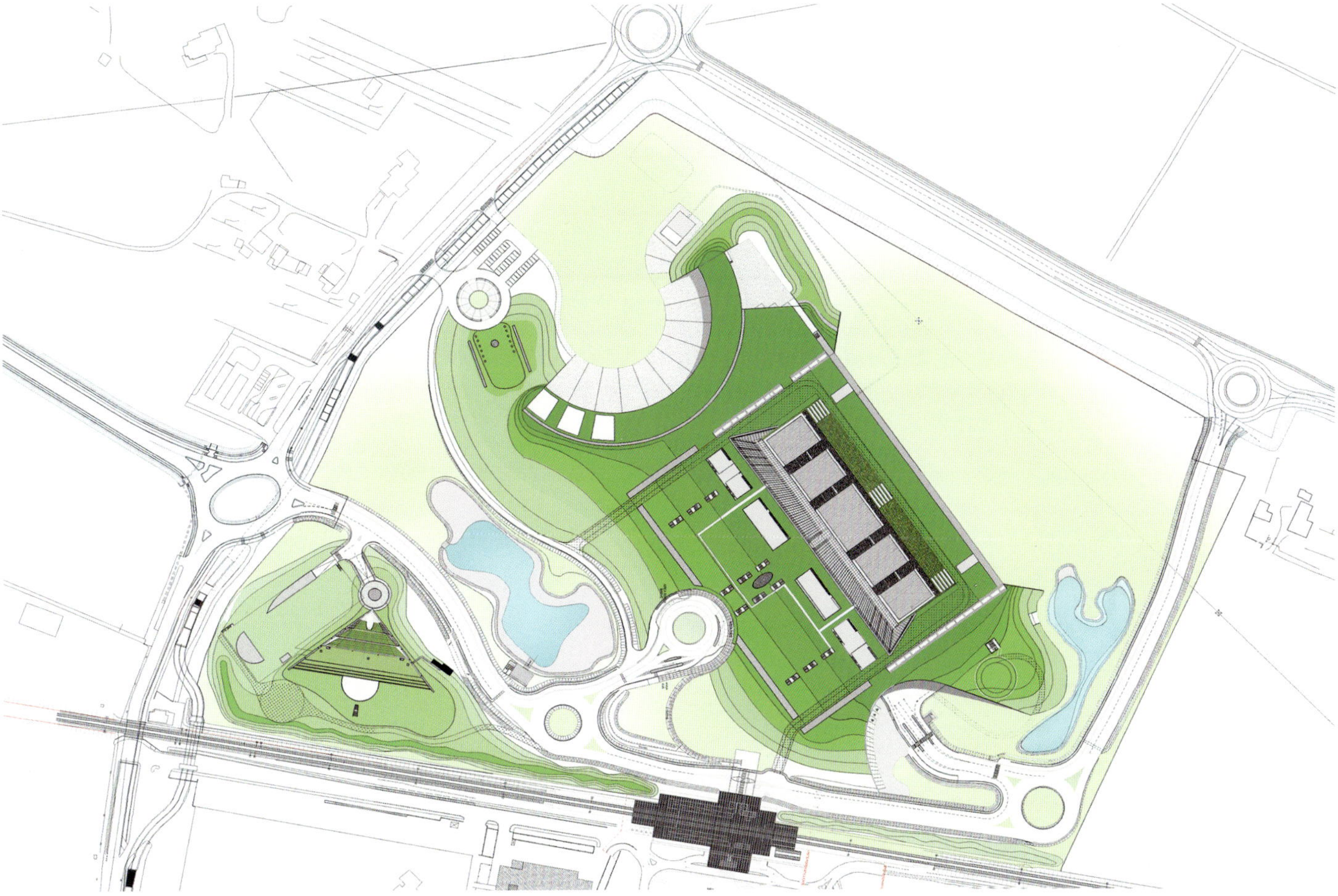

193

Banca degli Occhi

Venice-Mestre, Italy, 2009

This laboratory, under the aegis of the private foundation Fondazione Banca degli Occhi, is unique not only for the fact that it has more than thirty years of experience in eye transplants and training doctors in these techniques, but also that it has engaged in stem cell research with remarkable ophthalmologic results. The building, following a triangular plan, contains the stem cell research labs, a school for training professionals, operating and recovery rooms, and covered parking, as well as a large underground auditorium. The building is defined by two long trapezoidal walls, sheathed with a bronze patina finish and placed at right angles to each other; their projected tips are separated by a few inches, thus evoking Michelangelo's painting in the Sistine Chapel of God's finger transferring his élan vital to Adam. The roof of the building is a stepped section plane covered in fragrant greenery that can be appreciated in both an olfactory and visual way by the patients upon entering; it also serves as an open-air auditorium as well as providing an emergency exit for each floor. On the third side, the laboratory technicians and the patients have personal views to the plants growing outside their windows in wide earth-covered terraces. This building is located across the street from the Nuovo Ospedale di Venezia Mestre, recently baptized as the Ospedale dell'Angelo (The Angel's Hospital): the first "green" hospital in Europe.

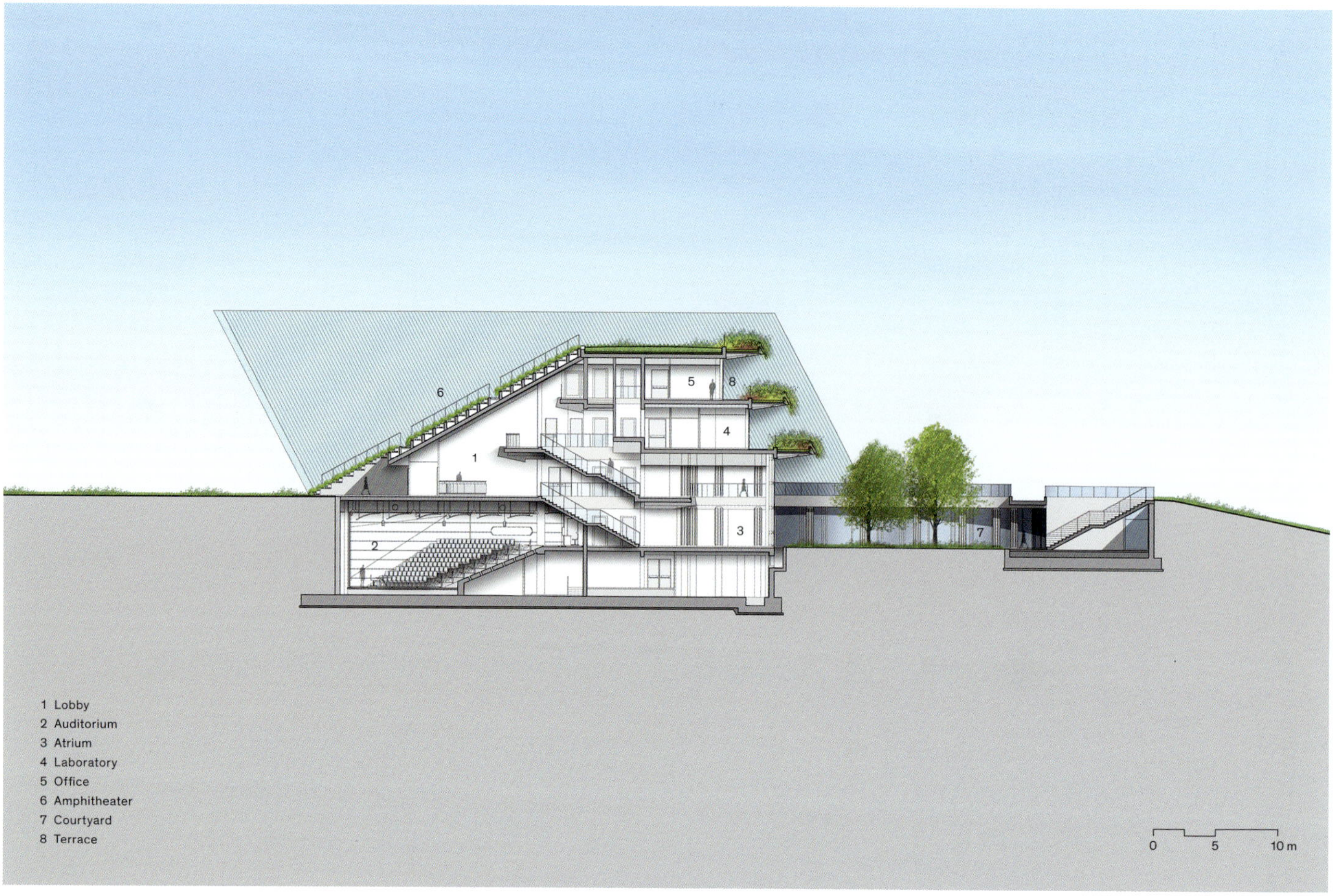
6
5
8
4
1
3
2
7
1 Lobby
2 Auditorium
3 Atrium
4 Laboratory
5 Office
6 Amphitheater
7 Courtyard
8 Terrace
0
5
10 m

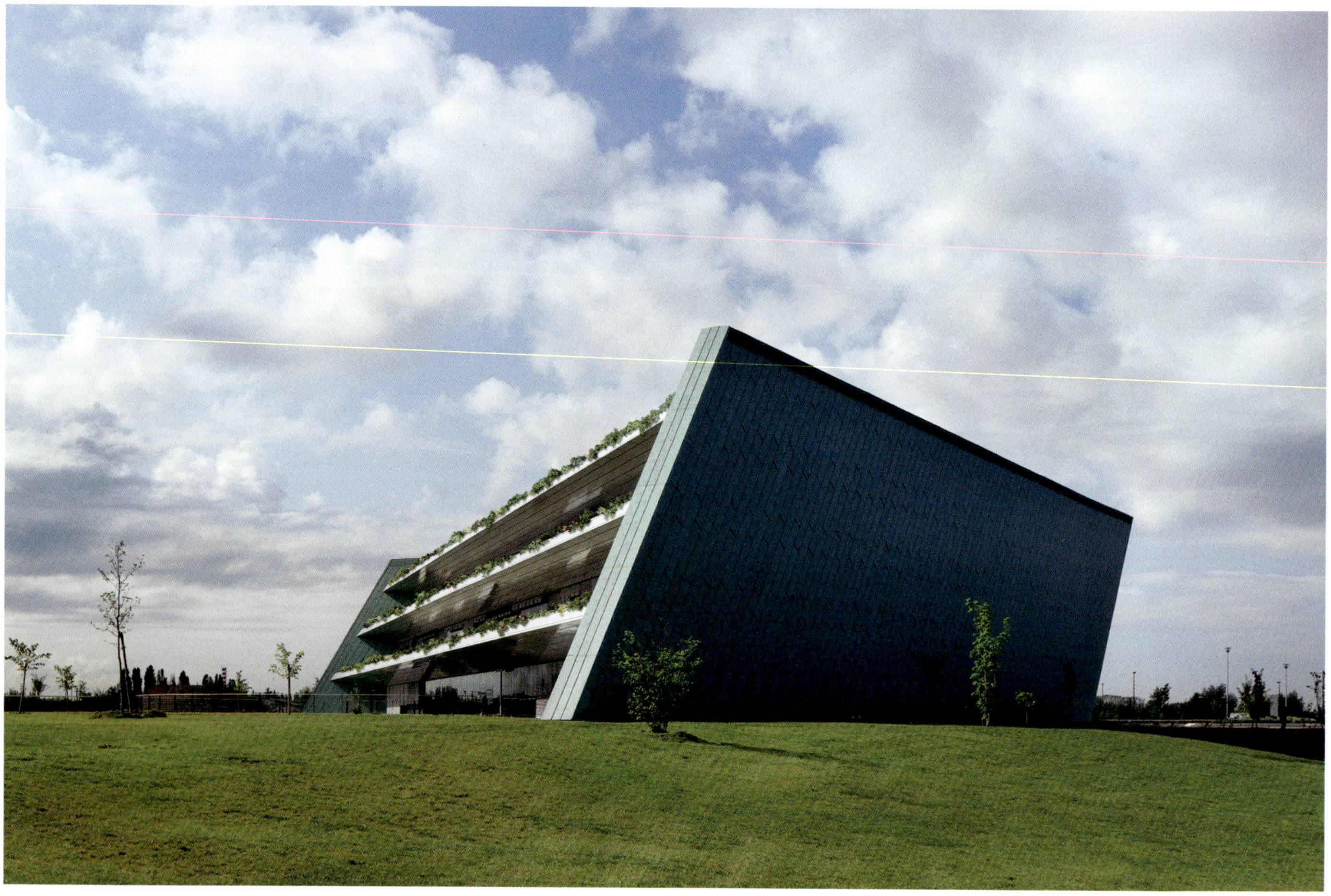

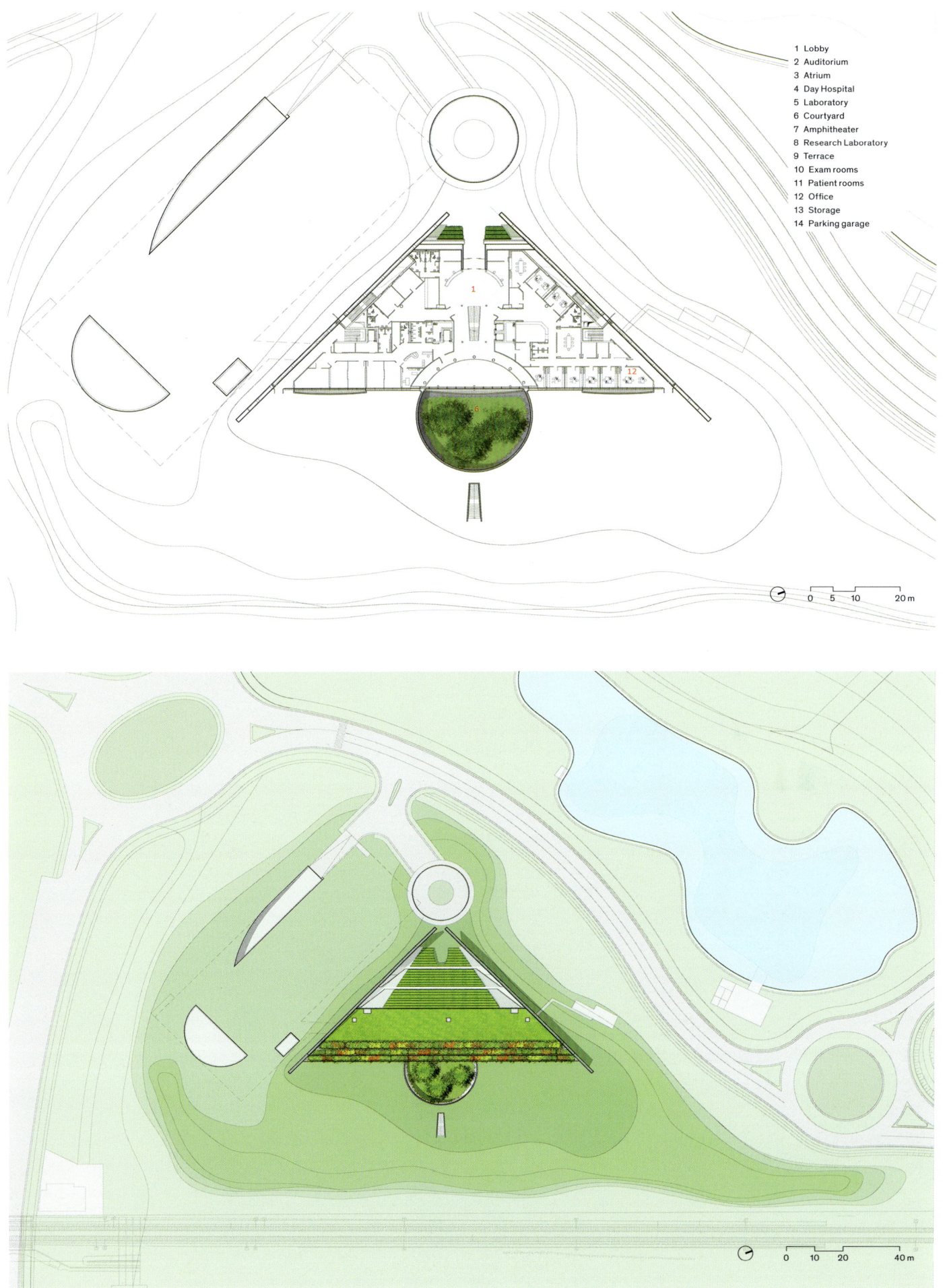

1 Lobby
2 Auditorium
3 Atrium
4 Day Hospital
5 Laboratory
6 Courtyard
7 Amphitheater
8 Research Laboratory
9 Terrace
10 Exam rooms
11 Patient rooms
12 Office
13 Storage
14 Parking garage

Ambasz's own semiautobiographical fable Designer/Producer *is the tale of a designer who, frustrated with the compartmentalization of design and with manufacturers' unwillingness to take risks, forms a cooperative with a model maker, mechanic, production engineer, and mold maker.*

Peter Hall

THE IMMORTAL
Peter Hall

Salomon saith, "There is no new thing upon the earth. So that as Plato had an imagination, that all knowledge is but remembrance; so Salomon giveth his sentence, that all novelty is but oblivion."
Francis Bacon, *Essays*

When I came out of the last cellar, I found him at the mouth of the cave. He was stretched out on the sand, where he was tracing clumsily then erasing a string of signs that, like the letters in our dreams, seem on the verge of being understood and then dissolve. [...]

Everything was elucidated for me that day. The Troglodytes were the Immortals; the riverlet of sandy water the River sought by the horseman. As for the city whose renown had spread as far as the Ganges, it was some nine centuries since the Immortals had razed it. With the relics of its ruins they erected, in the same place, the mad city I had traversed: a kind of parody or inversion and also temple of the irrational Gods who govern the world and of whom we know nothing, save that they do not resemble man. This establishment was the last symbol to which the Immortals condescended. It marks a stage at which, judging that all undertakings are in vain, they determined to live in thought, in pure speculation.
Jorge Luis Borges, "The Immortal"

A great deal of industrial design and architecture might be viewed, after Borges, as the vain pursuit of immortality. Our impulse to leave palimpsestic traces and mad cities to our offspring is periodically undermined by our suspicion that all novelty is oblivion, as Bacon reminds us. If the troglodytes of *The Immortal* found refuge from this paradox in the solace of thought and "pure speculation" (as, indeed, did Borges, the writer, librarian, and voracious reader), the industrial designer faces a more troubling impasse. Historically, designers have been shackled to the idea that they must produce to survive. In the United States, as has been well documented, the role of the industrial designer was first defined in the 1920s as a subsidiary of advertising, to sell consumer goods through the addition of novelty, of seductive styling. "The industrial designer began as the man who persuaded industry to make those dreary household gadgets and appliances look glamorous," noted George Nelson archly in a *Fortune* magazine editorial of 1934, "thus starting a love affair with the American housewife that is not yet over."[1] The tryst is still ongoing. The annual design issue of the *New York Times Magazine,* despite its best intentions to prove that design is not an "affectation or afterthought," spends most of its pages actually reinforcing the point, displaying designer jeans, $20,000 phones, and *Wallpaper* magazine, which, as critic Stephen Bayley put it, presents design as "meretricious exclusivity."[2]

The industrial design work of Emilio Ambasz belongs to a critical trajectory that runs in opposition to this perceived role of design as novelty, affectation, and afterthought, a resistant strain whose notables include Nelson, R. Buckminster Fuller, and the current crop of young Dutch designers associated with Droog Design, who have turned design into defiance of built-in obsolescence. Ambasz's studio generates most of its own projects, surveying the Borgesian impasse, as it were, from the caves of pure speculation. "That thing which does not come into being does not die," as the Zen Buddhist teaching goes. According to one insider who worked in the Ambasz camp during the 1980s, for every product that saw the light of day, there were dozens that remained only as thoughts. But when speculation is taken into production, as is the inevitable obligation of every designer operating outside academia, Ambasz's objects strive to serve human needs over desires. They aim toward what Victor Papanek, scourge of the design establishment, called in 1971 "honest design (design-in-use versus design-in-sales.)"[3] Papanek famously characterized design's pioneers—Van Doren, Bel Geddes, Deskey, and Teague—as window display and stage designers who brought visual excitement but not nourishment to manufacturing during the Depression, much as "the swollen belly of a child suffering from malnutrition gives it the appearance of being well fed."[4] Finding more to applaud in the wartime era of design driven by performance criteria and limited materials, Papanek called for designers to make greater sacrifices, and much more innovative work that would contribute to "real human and social needs."[5]

Ambasz's Vertebra chair, designed with Giancarlo Piretti, would appear to be a prize example of what Papanek was calling for. The first automatic, articulated task chair, Vertebra sought to make the device we sit in for hours a day less unhealthy. Or, as Mario Bellini described it, Vertebra strove "to make it possible for the user to consider the chair as a dynamic and active entity, changing its configuration automatically, whenever the body desires."[6]

Aside from establishing a benchmark for ergonomic seating (a point to which we shall return), the Vertebra chair was singled out by one triumvirate of critics—Stephen Bayley, Philippe Garner, and

Deyan Sudjic — as evidence of a structural change in the manner in which design is practiced.[7] In a climate of "just-in-time" inventories and computerized manufacturing systems, the consultant designer could no longer distance himself from the dreary subject of sales returns or the challenge of modifying products in response to production or market demands:

> Ambasz no longer expects commissions to come to him, instead he develops his products himself. With his Vertebra chair of 1979, for instance, he sold a part of the future business to toolmakers, diecasters and upholsterers, persuading them to make tools and dies at no cost in anticipation of a share of the profits. He also carried out his own market projection, and so was able to approach the manufacturers, Castelli, with a fully costed program. The company therefore was able to undertake the manufacture of the chair, itself a design of some ingenuity, with virtually no risk because the designer had removed much of the uncertainty from the design process.[8]

In Ambasz's world, the designer becomes a figure who not only speculates on what might be, but follows through by pulling together a production team and taking on full responsibility for the costs and societal impact of that team's creations. Industrial design is thus detached from its moorings to advertising.

It would be misleading, however, to categorize the work of Ambasz as "honest design" in the Papanek mold. In addressing Ambasz's influences, and his influence on the profession, it is important to note three important forebears: Bucky Fuller, George Nelson, and Ettore Sottsass Jr. Fuller's utopian view of technology and tireless search for inexpensive, mobile, resource-saving structures established the foundations on which designers like Ambasz would later build. Nelson's playful experimentalism with materials and processes, his deft visual wit and ability to recognize design's impact beyond the sales curve established an equally important precedent: in 1949, Nelson argued that the profession's most important job was to reintegrate a society shattered by the pressure of new technology, infusing "emotional content" into inanimate objects.[9] The Ball Clock, launched a year later, certainly succeeded in bringing cheerful modernity in through America's kitchen doors in its allusions to technological iconography — the atom and the asterisk. Nelson's position marked a departure from Fuller's version of design as a means by which technology is harnessed to man's physical and spiritual well-being. Nelson's demonstration that mass-produced perfection might be given a phenomenological twist, and that design might become the salve to — rather than the embodiment of — the march of technological progress had an important impact, in turn, on the young Ettore

Sottsass Jr., who worked in Nelson's New York studio in 1956. Sottsass subsequently recalled learning from Nelson that design is not just a matter of being creative or original, but "abandoning yourself" and "understanding where society is going."[10] In Sottsass and his Italian cohorts, designed objects became capable of framing questions.

Ambasz's pivotal contribution to the Italian-led antirationalist revolt was not as a designer but as curator of the Museum of Modern Art's landmark 1972 exhibition *Italy: The New Domestic Landscape*. The exhibition marked a significant change of perspective for MoMA, which, as Jonathan Woodham has noted, "had tended to focus on the aesthetics of the individual object or celebrated designer."[11] Featuring the work of Sottsass, Mario Bellini, Andrea Branzi, Gaetano Pesce, Enzo Mari, and others, *Italy: The New Domestic Landscape* gave gallery space to the Italian notion of objects as part of a utopian environment in which self-sustaining technology liberated the individual from the death spiral of work and conspicuous consumption, and with symbolism—as Sottsass stressed—launched him or her on a path of self-discovery.

Ambasz took a coolly detached, curatorial tone in his introduction to the exhibition catalog, identifying three prevalent attitudes toward design in Italy at the time. The "conformist" approach referred to designers who did not question the sociocultural context in which they worked, but continued to refine already established forms and functions.[12] The "reformist" approach referred to designers who had a "profound concern" for their role in a society that encourages consumption as a means of inducing happiness, and responded with a "rhetorical mode," producing ironic revivals or seeking refuge in natural forms. The third approach, argued Ambasz, was "one of contestation"—either refusing to take part in the socioindustrial system at all or engaging in "active critical participation." In this account, Ambasz began to reveal glimpses of the ground rules that would guide his design explorations in subsequent decades:

> To the traditional preoccupation with aesthetic objects, these contemporary designers have therefore added a concern for an aesthetic of the uses made of these objects. This holistic approach is manifested in the design of objects that are flexible in function, thus permitting multiple modes of use and arrangement. To one accustomed to dealing with finite shapes that can act as points of reference, such objects can be offensive, because they refuse to adopt a fixed shape or to serve as reference to anything.[13]

The statement might be taken to apply to Sottsass's subsequent Carlton Bookcase of 1981, perhaps better described as a labyrinth of shelves in search of a bookcase. Defiantly antifunctionalist, the

Carlton was a manifestation of Sottsass's argument that since there was no design solution that could not be replaced by another, one must initiate a methodology with the focus less on perfect form than on the "method of searching for form."[14] Yet the pursuit of objects that "refuse to adopt a fixed shape" applies more literally to Ambasz's Vertebra chair. Unlike the Carlton bookcase, the defiance of finite shapes inherent in the Vertebra's design has an eminently functional purpose: Ambasz claims that it emerged out of his oft-frowned-upon habit of swinging on the back two legs of a fixed chair. "I wanted it to move with me," he says.[15] As fellow task-chair designer Niels Diffrient has observed, this instinct is natural. "The more you lean back, the more you transfer weight on to the backrest instead of your spine." The Vertebra chair, noted Bellini, "does not engage the abdominal muscles, and allows the user to exercise the intervertebral discs, thereby actively maintaining the flowing of fluids in the dorsal spine."[16] Diffrient concludes: "It was a breakthrough product—it did what it intended to do with very few controls. The mechanisms were ingenious: It not only had a tilting back but a sliding seat coordinated with it."[17]

At the same time, one cannot ignore the semantic aspects of the chair's form. The now ubiquitous ribbed tubular treatment of the chair back's articulating mechanism clearly alluded to the spine, the soft flexing forms suggesting, without direct iconographic reference, that this servile seat might yet have its own soul. "Vertebra behaves like we do, as organisms in motion," noted one complementary text.[18] Subsequent Ambasz designs, as we shall see, play more explicitly with anthropomorphism, from the snail-shaped air filter to the dove-shaped water bottle. Design, as Ambasz has often stated, gives "poetic form to the pragmatic."[19]

After leaving MoMA, Ambasz found that green architecture was not enough to support him ("all my clients were other architects," he says[20]) so he turned—following the pattern of the young Italian architects he had feted at MoMA—to industrial design. In doing so, he developed a methodology—modernist in its follow-through, but with some of Sottsass's poetic flair—that established an important concession to our desire for traces. His objects aspired toward immateriality by recalling archetypes. "It has always been my deep belief that architecture and design are both mythmaking acts," begins Ambasz's fable "I Ask Myself," perhaps titled in an echo of Borges's essay "Borges and I." "I hold that their real tasks begin once functional and behavioral needs have been satisfied."[21]

If we were to borrow a literary mode to analyze the Ambasz oeuvre, then, we might appropriately turn to the archetypal criticism of Northrop Frye et al. Reading a text according to archetypal criticism depends on identifying themes, images, symbols, plots, and characters that correspond with recurring archetypes of myths and rituals. Its defiantly anti-Marxist premise is that these literary elements cannot be simply explained with reference to social, biological, or historical influences because they are linked to sources prior to these contexts. Such an approach, though criticized as reductivist, can in skilled hands be expansive, connecting diverse and various mythologies, enlightening the reader with the links, implications, and sometimes ambiguities embedded in a text. Borges, whose work encourages intertextual readings, offers clues to suggest that his layered topology of textual signposts might ultimately lead to a core of metasigns, or archetypes. He once argued that there are only four basic devices of all fantastic literature: the work within a work, the voyage in time, the contamination of reality by a dream, and the double.

Ambasz declares, consciously recalling a remark from Borges, that he is a "man of few ideas many times reformulated."[22] The ergonomic chair, for example, has remained a perennial pursuit at Ambasz's studio. The chair has a status in mythology that might at first seem contradictory to contemporary perceptions of ergonomic seating. In pre-Hispanic Colombia, for example, the most common portrayal of the human form was in a seated posture. Male deities, priests, and shamans were commonly depicted in ceramics and gold, seated on benches; their demeanor was invariably impassive, suggesting inner calm and contemplation, seated at the confluence of sky and water. In Asiatic countries, the seat similarly expressed synthesis, stability, and unity. Yet contemporary understanding of human factors suggests that the body does not thrive from prolonged stasis and stability, and the best chair does not remain fixed in a "comfortable" position but moves with the body. Considered in a mobile form, as a chariot, however, the seat's symbolism gains an interesting additional meaning. J. E. Cirlot observes the following characteristics of the chariot archetype:

> The charioteer represents the *self* of Jungian psychology; the chariot the human body and also thought in its transitory aspects relative to things terrestrial; the horses are the life force; and the reins denote intelligence and willpower.[23]

According to an archetypal reading, then, the chair in flux—the task chair—can represent both a position of power and inner calm as well as an embodiment of the human form. The division between physical seat and sitter is dissolved, the whole representing the body and mind in cohesion.

Ambasz applied the notion of the integrated sitter and seat to even the humble institutional chair, first with Dorsal (1978) and then with the basic office-worker chair Lumb-R (1981). Both incorporated a responsive backrest and contoured forms designed to maximize blood circulation and provide optimal weight distribution, but using simple flexing mechanisms. An archetypal reading might also identify in the chairs' profiles the presence of hybrid marks derived from the Greek alphabet, reminding us of the troglodyte's palimpsestic tracings and erasings in "The Immortal." The same elemental, delineated approach is apparent in the later VoX Contract Chair (1996) and the Stacker Contract Chair (1998), both folding and stackable institutional chairs that employ the letter *X* as legs and the structural means of support for the backrest, seat beam, and optional armrest and writing tablet. Again, a simple flexing backrest, made in Delrin (VoX) and ABS (Stacker), aims to provide better support for sitters in public places, who may find themselves spending up to three or four hours in a chair in a school, a waiting *area* or a factory. Symbolically, the *X*, at least according to the researcher of symbols Harold Bayley, denotes the union of two worlds—the superior and inferior.[24] It would seem a perfectly appropriate letter to support the union of mind, body, and man-made environment.

The 1991 Vertair chair pursues more anthropomorphic associations in an upholstery system, a series of narrow, overlapping leather bands stitched to elastic, all expanding and contracting with the sitter's motion. Unlike the recent spate of hard-edged, airily named high-tech ergonomic chairs that have emerged from the furniture giants and that boast dozens of operational levers and tension controls, Vertair's automatic adjustment does not require the office worker to read and digest a hefty instruction manual before sitting down. Levers, buttons, and knobs are banished from the form in favor of a minimalist form Ambasz calls "soft-tech," but might equally be described as crustacean. Allusions to animals as life forces are common enough in mythology; in Ambasz's designs, the reference recurs as a playful hint.

In three subsequent designs—Brief (1995), Max Operative (1996), and Tennis (1997)—the seating system gains a more sail-like quality, becoming the winged chariot, perhaps. Formal origins might be traced to the earlier Qualis seating system (1989), which poses the question "what is it?" through variations on a theme: high-tech features, such as a self-adjusting tilt-forward to relieve pressure on the thighs, are underplayed in favor of a soft, folded rectangular form gently hinting, perhaps, at Magistretti's Sinbad chair (inspired by a thrown blanket). The soft geometry is then applied to various configurations, with arms, without, with wheels and without (again, recalling the inscribed letter). With Brief, Max, and Tennis, however, ergonomic issues are more thoroughly addressed. Office chairs, until recently, were designed for the 50th percentile man, a somewhat mythical figure in himself, vastly outnumbered in offices these days by men and women of various shapes and sizes. One solution to the diversity of sizes in the workplace is to manufacture chairs in three sizes — small, medium, and large. The downside is that office workers are less likely than ever to stay in the same job for more than a few years, increasing the likelihood that he or she will bequeath an ill-fitting seat to his or her replacement. Ambasz's solution: a height-adjustable backrest and armrests, to provide lumbar support for a variety of sizes. The function is visually expressed in the sail-like forms of the backrests, which convey their independent maneuverability.

If the Ambasz task chairs are an extension, to paraphrase Marshall McLuhan, of the entire human form, then his pens are an extension of the hand. Ambasz first explored this contention, or that "everything is a prosthetic device," as he puts it, in a collection of durable, brightly colored ballpoint pens designed for schoolchildren.[25] The personal challenge was to create a pen that doesn't break in the pocket when you sit on it. Ambasz's solution was a polypropylene form which, like the human finger, transforms from flexible (for carrying) to rigid (for writing). A central ribbed midsection on the pen shaft flexes with movement, but the tubular lower and upper housings on each side can be twisted and snapped together to form a rigid shell that encloses the center. The upper housing also serves to protect the nib when the pen is in its "flexible" mode. Designer Eric Chan, who assisted Ambasz on the project, notes: "It's a very poetic idea, with a distinct before and after. Emilio always wants to simplify a mechanism, not in a mechanical way, but in an organic way. He's very good at nontraditional ways of solving problems."[26] To underline the point, Ambasz's later Magic Wand Automatic Roller Pen (1998) eliminates even the act of removing a pen cap or depressing a button to begin writing. Using a hidden automatic mechanism, the Magic Wand only proffers a writing nib when the user takes the pen out to write. As Chan puts it, "For Emilio an idea has to be groundbreaking or it's not worth doing."[27]

The literary critic James E. Irby has written of Borges that he demonstrates the "magic in obtaining the most powerful effects with a strict economy of means."[28] Sottsass has extended a similar notion to Ambasz's design, arguing in 1988 that Ambasz's body of work is "born from an obsessive search for primary principles, from a careful and wise observation of the surrounding reality, from an identification of humanistic problems."[29] Ambasz's various lighting projects are the result of a search for a primary geometrical principle that might maximize the degrees of freedom available to a cylindrical form. The first manifestation, the Polyphemus flashlight (1983), demonstrates the answer: an elliptical cylinder is sectioned at a forty-five-degree angle, producing a highly maneuverable one-eyed head. When the top half is rotated, the light forms an L-shape that is easier to carry and more direct than the conventional cylinder flashlight. The beam can also be directed with the light in a standing position or, using a magnet in the light base, attached to any metal surface. Characteristically, the one-eyed light is named after the Cyclops who was blinded by Odysseus on the island of Sicily.

Clearly, there is certain magician's delight inherent in developing this array of elemental forms that do not reveal themselves as lights or pens until used. In this respect, Ambasz shares something in common with a peer whose work he greatly admires, Richard Sapper. Sapper exemplified the transformational product in a series of IBM ThinkPads, which expand out of slick black cigar-box-sized volumes. Indeed, Ambasz produced a version of the IBM computer that extended the notion to a user-determined modular system. The IBM Portable Desktop (2000) proposed a one-size-fits-all smoked, tinted acrylic container, out of which would spring, as per the user's requirements, a height-adjustable screen (twelve- or sixteen-inch), a laptop-sized CPU, a DVD player, a wireless keyboard, and a handheld device/GSM phone. Again, a literary parallel for the transformative container can be found in James E. Irby's account of Borges, who "uses mystery and the surprise effect in literature to achieve that sacred astonishment at the universe."[30] When the black Pandora's box contains a customizable computer—a window to the mysteries of the World Wide Web—sacred astonishment might admittedly be more akin to the horror of Joseph Conrad's *Heart of Darkness*.

More slicing of geometric volumes produced Oseris, a range of low-voltage spotlights named after the Egyptian god Osiris, who was murdered by his brother and sliced into fourteen pieces that were distributed around Egypt. Extending the principle behind Polyphemus, Ambasz established that if a semispherical volume is cut twice (if not fourteen times) with a plane, the intersections generate perfect circles, which, when turned against each other, describe a movement from zero to ninety degrees. The light can also turn on its vertical axis, providing a high degree of maneuverability that can be orchestrated in museum and other exhibition settings with the help of a printed scale.

Most ambitiously, Ambasz's Soft Series designs for handheld consumer electronics foreshadowed the much-ballyhooed ubiquitous computing age into which we are rapidly being thrust. With the soft Portable Radio/Cassette, Notebook Computer, Handkerchief TV, and Phone—which encased all the mechanical and electrical components inside padded leather skins—Ambasz made our communications and entertainment devices "wearable." The Handkerchief TV, like Sapper's ThinkPads, only reveals its function when unfolded like a piece of origami; in folded form it resembles a wallet. Each leather plane is opened to expose a distinct function: screen, antenna, battery/

speaker, and external ports. But where Sapper celebrates the engineered mechanisms of his unfolding designs, Ambasz pursues simpler, more organic guises. The very human, ceremonial aspect of unfolding a soft leather case to reveal a phone is ultimately more important to Ambasz as a ritual than showcasing the mechanical feats required to create it.

As was noted in the catalog of Ambasz's 1992 exhibition at the Institute of Contemporary Art at Tokyo Station, "The miniaturization of technology has led to the increasing portability of televisions, computers, cassette tape players, and other commonly used products, but the forms and materials of these products have not fundamentally altered to reflect this technological development." Indeed, many of the forms proposed for these consumer electronics goods were derived from their mechanical forebears—the anachronistic typewriter interface as a means of inputting data into computers being the prime example. The Soft Series was a response to this perceived lack of momentum in design, positing a "zoological species" in place of hard-edged products with forms derived from their former, mechanical forebears.[32] The seeds of this foray into metaphor, however, can be traced back to Mario Bellini's "zoomorphic" Divisumma 18 calculator of 1970, which sported soft, nipplelike buttons and a rubbery, seal-like surface. Today, the old mechanical tropes are increasingly vanishing. Boomboxes have since 1992 mutated into an endless array of personages, from submarines to jeeps to compacts; cellphones are "clamshells" or—increasingly—resemble hearing aids; and cassette tape players have become miniscule MP3 players resembling bars of soap or jewelry.

Perhaps more than heralding the plunge of industrial design into metaphor, the Soft Series anticipated the eventual absorption of electronic devices into the fabric of the built environment. In this respect, the immortality/materiality paradox posed at the beginning of this essay is compounded when we consider what happens to the industrial designer when a manufacturing-based economy turns into a New Economy and then a Now Economy. The implication might seem to be that the designer is left without a job to do, without an opportunity to leave a trace for posterity. "The visual design of objects and places becomes less relevant," writes Ole Bouman in *Archis* magazine, "if those objects and places, the good old *A*s and *B*s, are displaced by the movement between them as the criterion of our existence."[33]

One of the most beguiling recent Ambasz studio projects, the Saturno Street/Highway lighting, would appear to embrace this critical shift of attention away from objects and toward flows. The Saturno light not only provides a design solution to a problem of infrastructure, the highway, or what Marc Augé would call a "nonplace,"[34] it aspires toward a form of "almost transparent presence," as Ambasz

The Western notion of Man's creations as distinct and separate entities—in contrast to those of Nature—has exhausted its intellectual and ethical capital. An emerging man-made garden is overtaking the one we were originally given. We must create an atectonic notion of architecture, where architecture is conceived as an integral component of that emerging man-made Nature we are willingly, as well as unwittingly, creating. I see the task of the architect to be that of reconciling our man-made Nature with the organic one we have been given. EA

Industrial and Graphic Design

Emilio Ambasz's industrial and graphic design accomplishments are of immense scope and astounding variety. They range from diesel engines, and street-, flash-, and spotlights, to portable TV players, flexible pens, fold-out watches, and innovative office chairs. To mention one project in particular, the Vertebra Chair came to be known as the first office chair that automatically adjusted to the occupant's body. By placing ergonomics alongside aesthetics, the chair introduced a new way of thinking about office furniture, and embodied Ambasz's striving for design as an extension of the body. By first designing and engineering products for himself, without a client's commission, and by building prototypes as well as the machinery required for the manufacturing process, Ambasz was able to offer finished and ready-to-go products, thus reducing both production times and costs, and extending product durability. Ambasz's graphic design of posters, logotypes, and books shows a continuous ambition to go beyond the two-dimensional limits of paper and image—as seen in posters for the Geigy exhibition, the corporate identity of Mycal Group, and others.

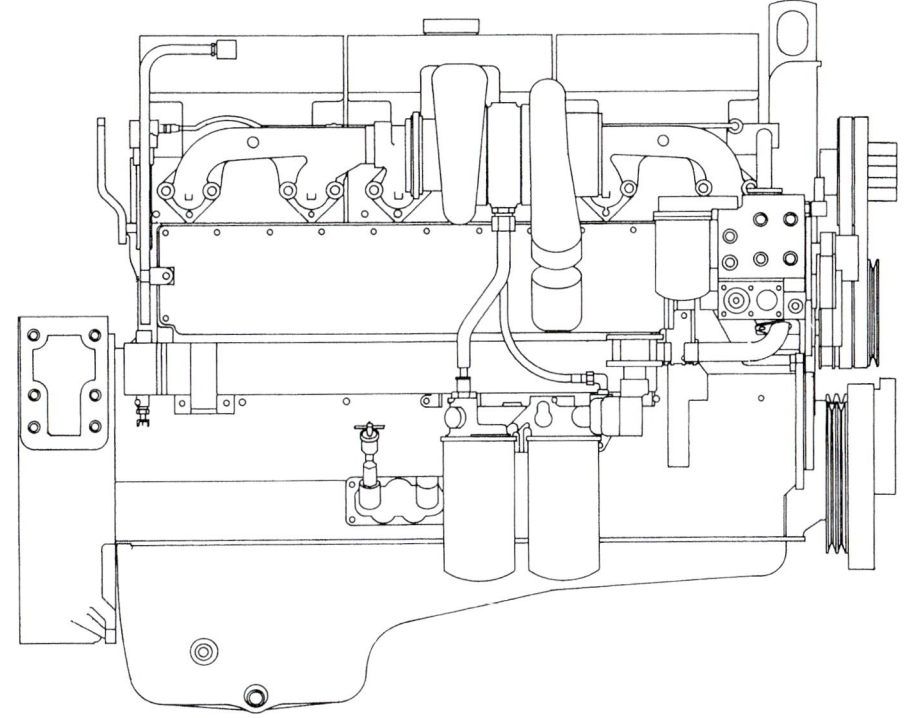
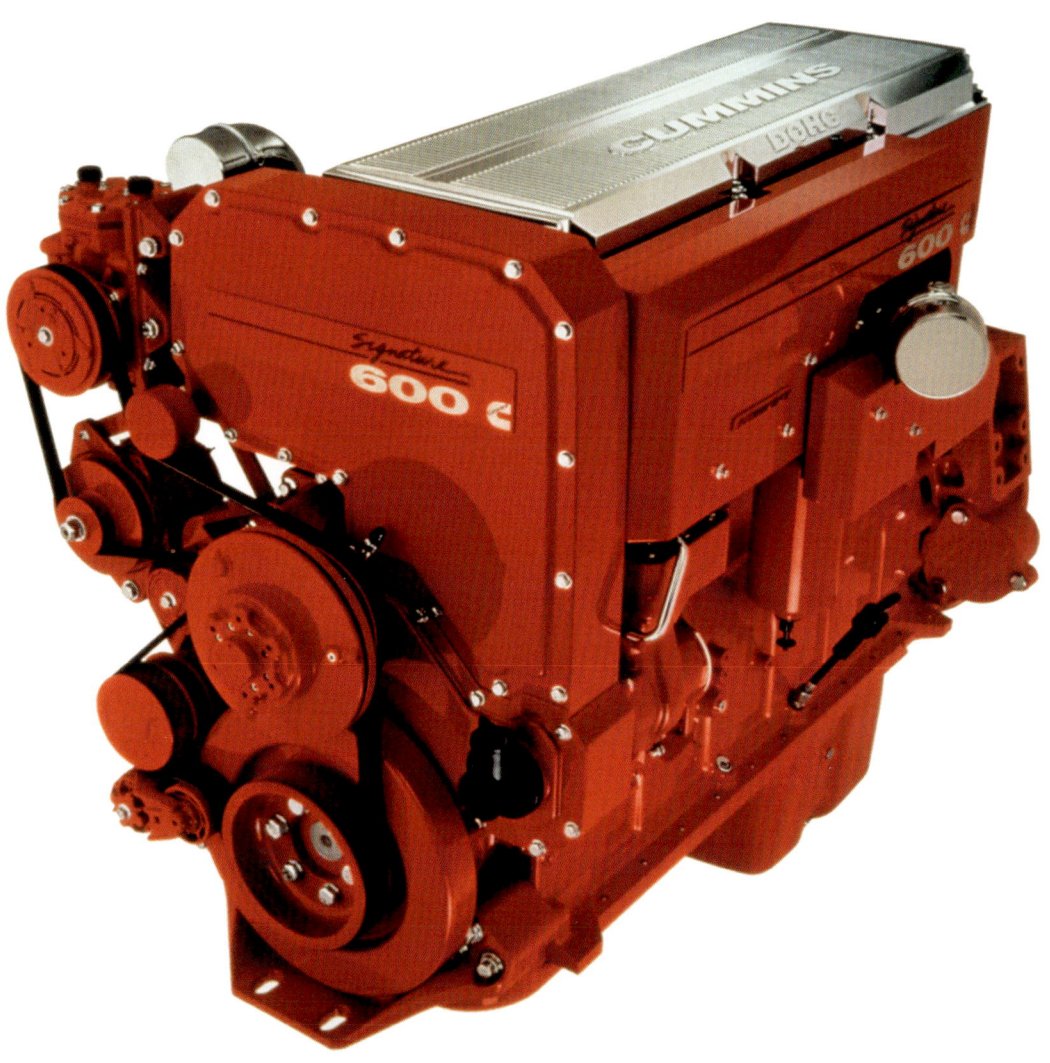

Cummins Diesel Engine, 1982

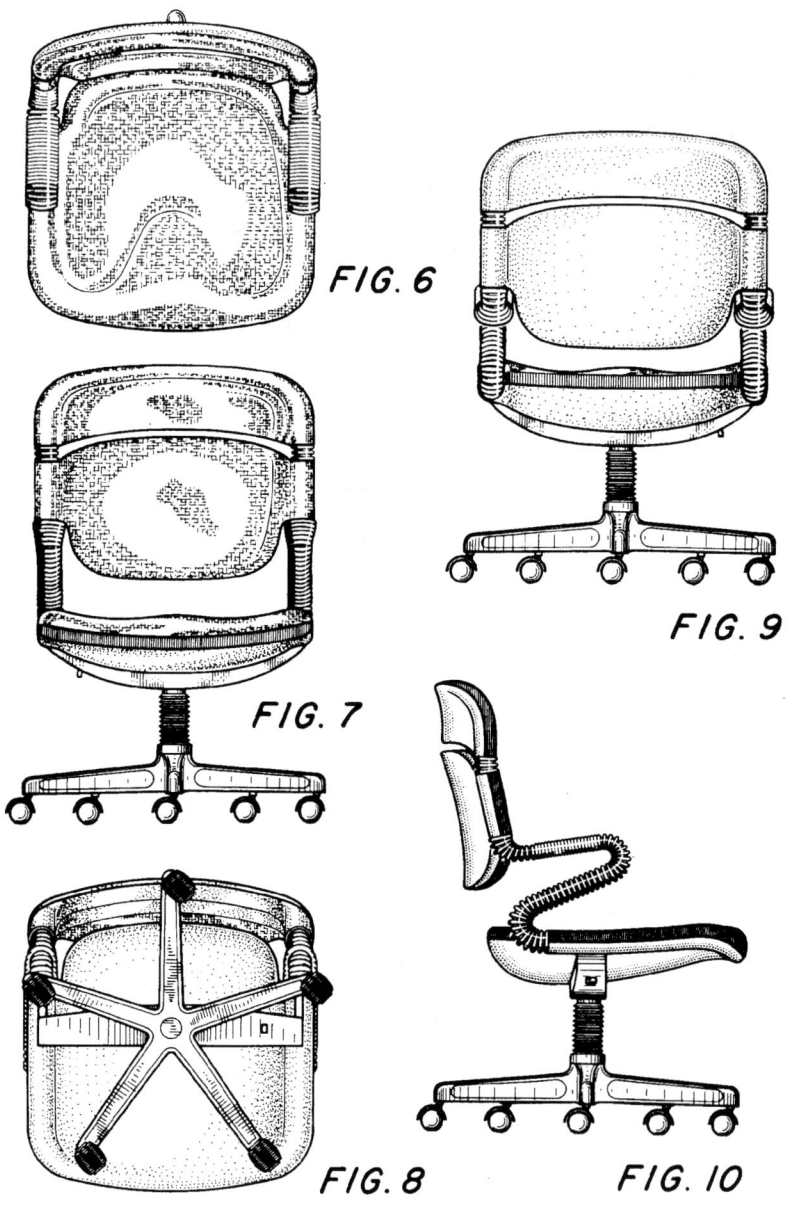

Tennis Office Seating, 1997

Vertair Chair, 1991

Dorsal Office Seating, 1978

Vertebra Seating System, 1974–75

VoX Contract Chair, 1996

Industrial and Graphic Design

Saturno Street / Highway Lighting, 1998

Industrial and Graphic Design

228

Agamennone Light, 1985 >

Polyphemus Flashlight, 1983

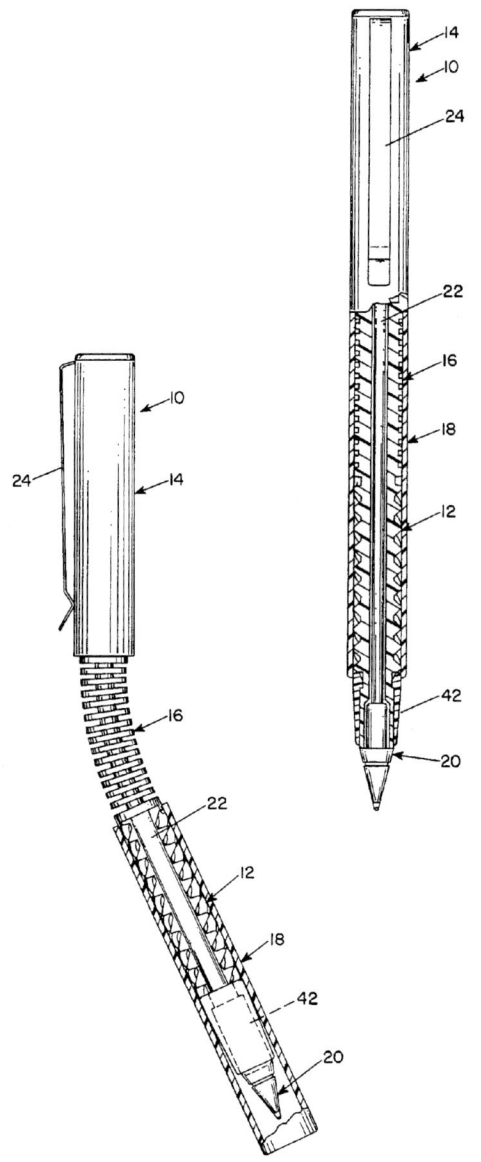

Flexibol Pens, 1985

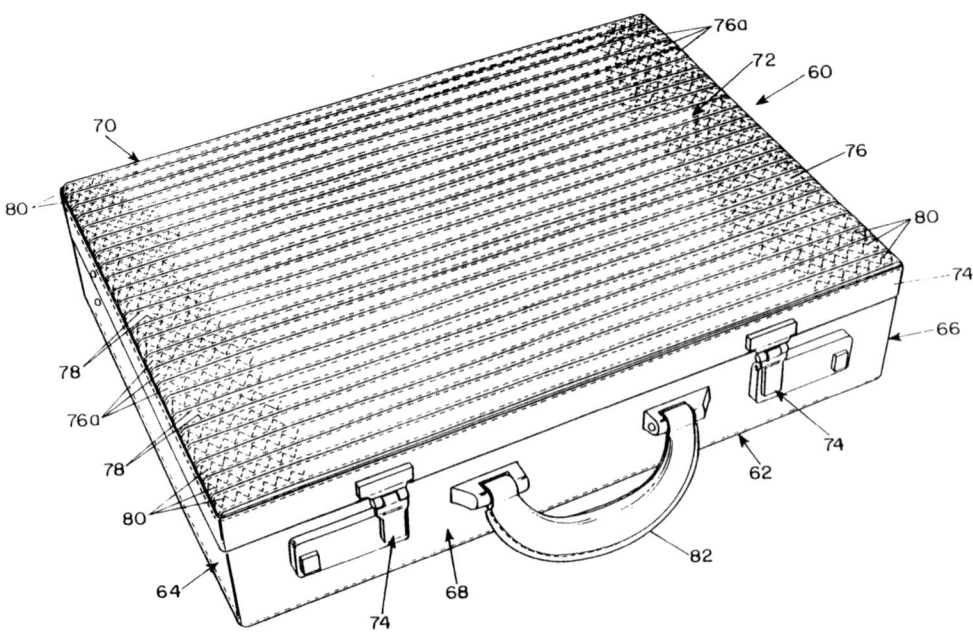

Logotec Light, 1984

Oseris Spotlight, 1985

X-Pand Briefcase, 1985

Handkerchief TV, 1990

Soft Portable Radio / Cassette Player, 1990

Geigy Graphics Posters, 1966

If one finds the quintessence of a problem, one will have better access to an irreducible solution. The thread supporting my design quest in every area—my products and my architecture—is a single preoccupation: finding the root of the problem, its essence.

As for expressive means, I seek to approach a design problem in the most crystalline, austere, and graceful manner. I long for an architecture which has been reduced to essentials and which, at the same time, is an architecture full of potential meanings. Such concision is the method by which to achieve a multidimensional, epigrammatic architecture. EA

< Residence-au-Lac, Lugano, Switzerland, 1983

INTERVIEW

Hans Ulrich Obrist with Emilio Ambasz

HUO I would really like to focus the interview on your work as a curator in the first chapter; I wanted to start with this. You said something at the beginning of this interview which is somehow related to that question, it is actually a notion that also appears in your writing. You have just been saying that you don't engage in architecture criticism but write fables. I was wondering if you could tell me a little about this practice of writing, is it a daily practice or is it only on occasions? And what is your mode of linking architecture to literature? It is something which is a bit forgotten now, strangely, but I have in mind Orhan Pamuk, the great Turkish novelist. Last month, he said that if you look at Istanbul, you may fancy that Istanbul was maybe invented by Gérard de Nerval or, in that vein, we could also say that Italo Calvino was a visionary urbanist.
EA They are both very important figures, but we should not forget Baudelaire.

I was wondering about the whole idea of literature and urbanism because, strangely now, we have all these links between contemporary art and architecture, between design and architecture, between architecture and fashion (I am thinking about all the flagship stores) so we have a lot of those bridges, but one of the most important things, which is the link to literature, has kind of faded away. It's the same for contemporary art. With the Surrealists there was always a link to literature, as with all the twentieth-century movements, but now it has ended really with the nouveau roman; Alain Robbe-Grillet was the last one who did it, somehow. So, that whole complex of your relation to literature and your producing literature interest me greatly.
Basically, I don't write theory because I believe that theory deteriorates very fast while fables do not. I write fables because I think that they have the power to grip the heart of the reader, if they are any good. I am all for seducing; that's what attracts me. I also use it in a certain way because by means of a fable one can, in some subtle way, if one understands its deeper meaning, or better saying, if one gains an insight into a fable's essence, we can use that original insight as a key to understand other mysteries. That's why I wrote fables. I call them *Working Fables: A Collection of Design Tales for Skeptic Children.* They are eminently design tales; some of them have morals, some of them are parables, some of them are just paradoxes. In addition, I also like writing on behalf of other architects. I take great pleasure in celebrating their creations. I had the greatest pleasure, as I told you before, in writing for James Sterling on his Stuttgart museum. I wrote for Tadao Ando. I only care to celebrate; I don't have a great interest in doing criticism. It is not that I think it does not merit to be done, it's just that I am a hedonist, and would rather seek to take pleasure in what I do. I like to sing the work of others, even if their own song is very different from my work. For example, I wrote an article on Frank Gehry's chair designs. I flatter myself thinking it was a good piece. I know he liked it. This gave me great pleasure. But the point I should make is that I do not know what I am going to write before I sit down to write it. I do not write every day. When I have to write it sits in my head for a while and then maybe I get one or two ideas. I never want to write these down. When I sit down to write, I usually dictate them into a little tape recorder, just like yours here, with the commas, with the periods and with everything it needs to stand as a well-drafted piece. As I am dictating my voice changes, it becomes deeper, resonant; it is almost as if I were seeking to seduce or charm someone. Only at the end of the recording do I gain cognition of what I wanted to say. Sometimes it is good, sometimes negligible. But then, I can always throw away the tape and start again.

That's interesting. So it's not a method of writing, it's a method of speaking.
Yes. Because to me, what is lost also is the fact that one should read aloud literature for the sound of it. I think that's the way literature started; as a parliament. I was once very moved by something that Derek Walcott, the Antillean poet, said about the Russian poet Marina Tsvetaeva. Of course, I don't know Russian, so I can't read it, but he was explaining that pronouncing the Russian word for "moon" forces the lips to adopt the shape of the letter O, the shape of the Moon. I found it was an extremely beautiful image. I always have the suspicion that the mouth is, for a British speaker, perhaps one of their areas of sensorial pleasure they allow themselves.

Anyway, to come back to the subject of your question about writing, MoMA has just published *The Universitas Project, Solutions for a Post-Technological Society,* a project: I produced at MoMA only thirty-five years ago, and I had to write the introduction. It had to be clear, technically descriptive of its purpose, and list its propositional points. To me it was like walking up a mountain. Ask me to write something which is not descriptive of something that happened before, but something that in some way suggests something which one wishes to happen in the future, and that I will do with great pleasure.

Your concern is literature. But I am not a man educated in literature or anything of that sort, I am just a reader. I don't read critically. I don't use a method for that. I dislike immensely having to read something which is stylistically badly formed. As you know, my original language is Spanish, not English, but I don't write in Spanish because Spanish grammar has such a strong structure that one starts saying one thing and one ends up saying something completely different because the structure of the sentence must be obeyed. In English I can manipulate it much better, I feel much freer, so I always write in English. It's the only language in which I care to write.

That's interesting. I am Swiss and Swiss German is not really a written language so in Swiss literature, if it's Dürrenmatt or if it's Frisch.
If I recall correctly Frisch was a Swiss architect, playwright and novelist. Durrenmatt was also Swiss.

Yes, they have always been in some form of exile because they speak in a language which is not their own. A Swiss writer writes in a language that is not Swiss. So this whole idea of being an exile language writer is something which then, for you, matters?
It does, indeed. Imagine my satisfaction when I discovered that Joseph Conrad's mother tongue was Polish. You cannot find a better stylist.

I would be interested, without pinning you down on some influences; I would be interested in seeing in terms of literature who have been, for you, the writers who have somehow inspired you. Would you really take to Orhan Pamuk's idea that in terms of conceiving cities and inventing them, literature is key?
Pamuk is very right and probably in the back of his mind so was Baudelaire with the *flâneur*. He gave us a most suggestive optic to perceive Paris. Later, the one that celebrated this city beautifully, is Benjamin. You see, in phenomenological terms, it is very, very hard to perceive a city unless you fly over it or at least read maps, and many people cannot read maps. But if you have acquired a literary mode of perceiving it gives you a filter as well as a frame of reference so that you can see a fragment of the city here, and a fragment of a city there, and doing so may move you or upset you, but it provides a certain way in which you can accommodate it in your mind. So the phenomenon of the emerging city, which is, of course, the phenomenon of a new urban jungle, can be in some way assimilated without too much fear. A jungle always gives fear. By such literary devices the urban jungle can be modulated and given room to roam. I have always been fascinated by Colin Rowe and his idea of the collage city, because he had a way of describing the city by dissimilar aggregates. For example, when I did one of the projects I called the Chicano winery, a Mexican-American project; I didn't want to design it with a pencil. I wanted to write it first and from the writing then asked different people to make me drawings of how they imagined it looking. I thought that one could, in said manner, be far more suggestive, in some cases, of an architectural idea than by actually imposing a drawn image. But, truly, I certainly have zero pretensions to any literary talent.

Influences I had, of course. Everybody says, "Oh, you are Argentine therefore your influence 'of course' is Borges." One has to say, "Borges, yes, but not for what they think." Borges is for me an inspiration just for writing very, very short sentences, very clear, and for never giving away a story with words. You have to guess it, as if it had a flavor: you have to smell it out, it's not in the bottle, nor in the liquid. That is what you learn from him, and that's the way one knows how to do it. He was, of course, immensely influential. Some people may say I am a minimalist. No, that is not correct; I am an essentialist. If I may paraphrase Paul Valéry: how was it he said it? Oh yes, "I want to be essential, not like a feather, but like a bird." I really strive for that.

You refer to a number of different authors; if one looks at your favorites, probably one of your own most quoted texts is the *Anthology for a Spatial Buenos Aires,* which is like a founding text; it almost has the character of a manifesto. I have two questions in relation to that: one is the notion of the manifesto. I have always been thinking that it is extremely strange that in our time not only do we have no more movements, but also we have no more manifestos. I was wondering how you felt about that, if you had ever felt in the sixties the necessity for manifestos or if you felt that that was somehow then already obsolete and how you feel about it today. Secondly, more specifically, I was curious if you could tell me a little bit about this *Anthology for a Spatial Buenos Aires,* which very much in an Orhan Pamuk way, forty years before, has actually kind of invented Buenos Aires in some way.
The anthology was part of my thesis when I had to get my master's degree. That was not required, it was something I thought one should do to provide a context for my master's thesis. My thesis was the design of the National Library of Buenos Aires, where Borges had been the director. Of course, being blind, he couldn't see my project; that gave me great courage! But the great omnipresence is Buenos Aires, a city where I wasn't born—I was born in Resistencia, which is twelve hundred kilometers to the north of Buenos Aires. Buenos Aires is a mental city and it is The City, as I said in the Anthology, that all the Argentines reinvent. It is also the longed-for city we would like it to be since we are so far away from any great cultural center. So Buenos Aires in part is Barcelona, in part it is Paris, in part it is Geneva; it has little fragments of each of those memories that immigrants, or people who travel want to bring back, as good cannibals, of that which has seduced them in their travels. But the anthology, I must say, is for me the proof that maybe I am not that very wrong because it is a fable; it is a collection of fragments of things written by others; by juxtaposing these fragments I generate an image of the city, which probably was never perceived in that way. Again, I would have been completely lost if I had to write a theory how to understand this city and give all the bibliographic references. I thought it was much better to do it in this manner. But to come back, was it a manifesto? No. It was a statement of melancholy, but a melancholy I do not want ever to be satisfied. I would be very frightened to go back to live in Buenos Aires; I want rather to inhabit it mentally.

That brings me to the manifestos. I think we are all aware of the errors and tragedies generated by architects wielding manifestos they, in good faith, think can change the world. Our generation, having seen utopias that have misled people in a journey through the desert for forty years only to see themselves back in Egypt, has grown quite skeptical of the overarching manifesto. We are a bit frightened of them. We have grown a bit more concerned with the journey of the twenty-four hours of the day. That's why I was so interested when Henri Lefebvre was writing about the sociology of the day in *La Vie quotidienne* and I invited him to participate in the *Universitas Project.* But at the same time, I can see that manifestos are an absolute necessity, almost an obligation, because they are metaphoric and therefore they are very concentrated. They have the power to instill an idea, sometimes even an idea and a half. The great disaster occurs when they contain two ideas!

The question is, how are the manifestos presented? Are they to be written, with the massive distrust we now have of words? Are they to be presented as images? I am a man who makes images because I have an immense distrust of words. I think that the supreme misfortune occurs in design or in inventing something when the word arrives before the image. I think I am not the first one to say it; I would like to think it is my original idea, but I know it's not! I design without words; I clean myself of words. Before designing I want to take a verbal laxative because words belong to a domain that is already semantic; it is already established, it has its limits. I prefer the image to come to me, if I can, without any inhibition. Once the image is created, and only then, do I try to understand what ideas are embedded into those images. I have written it and said it before. I believe that words operate effectively in the domain of established and accepted types; of course we don't forget that it is the destiny of types to become stereotypes. If I were to use words I would write poetry; to me everything is concentrated there and I can find everything I want to know by reading Lucretius's *De rerum natura.* Granted, of course, there was the school of Stoics preceding Lucretius, but for me in his poetry one finds an essential formula which is constantly giving, almost like a tree that gives seeds; another tree will emerge from that seed, but it will not be the same tree. Theory is the seed turned tree. For me all images that may be contained are within the seed. So I would not produce a manifesto in an essay format. I would fancy proposing an alternative mode of existence only through architectural images. At the same time, I would like to say immediately that when my slogan "The green over the gray" came to my mind, I was in truth writing a manifesto. Slogans have a great amount of suggestive power because people are seduced by words.

And which project expressed this?
"The green over the gray." That's it. The manifesto is in the title. I want very much to write a book made up of titles, of aphorisms.

Can you tell me about this manifesto, about the "green over the gray." You have written it down?
Plenty of times; but when I came to the moment of synthesizing it in that essential slogan—for it's a slogan—it rendered, *malgre moi-même,* an intellectual shortcut for the journalists. For example, when I had a show at the Triennale di Milano in May 2005, all the journalists were always onto the "green over the gray" because that is synthetic, it has a way of being grasped by everybody. As a slogan it is more astute than what I am, and it has the capacity of summarizing memorably many thousands of words. I believe that these slogans have much power, to guide as well as to mislead, so that is why I am now keeping away from fables and more into titles, into aphorisms. But again I insist they are ideas that came to me after the image was created. I didn't know what I wanted to say when I was designing.

Do you have a list of titles? Is it a long list?
It's not a list of titles because then it would be an index of titles as the book itself. So I could do it in one page.

It is a small book?
Oh yes, it has to be small and that brings us to the point that people do not read long things nowadays; they are educated into magazines and even magazines drape little boxes on a condensation of the main text in case people read only the box and not the article in the magazine. Is it wrong? Yes. Is it bad? No. If it's bad literature it is bad whether it's a 500-page book or a little short paragraph in a box. It depends on how gifted the writer is.

Now the question is where is the ambition for acceding to that plateau where it merits being called literature. That is probably lost, in grand part, because there are many rewards to he who writes articles for magazines, to he who makes snappy statements on television. I am quite often on television as well as on radio; because when they ask me a question I somehow manage to invent a memorable answer. Sometimes I reformulate what I have already said before, of course. But it is a risky gift and it frightens me because you can seduce and mislead people by that. So when I do that type of thing I try at the same time to show them that here with one hand I am catching your heart, while with the other one I want to reveal the trick to you so you are not the victim of any hunter who is not honest enough to tell you how to escape from his net.

Sartre always talked about this idea of being a public intellectual. You have mentioned the use of television and mass media and there is today an increased presence of architects being public intellectuals. If there is no one after Foucault and Deleuze there are no longer so many philosophers assuming that role of public intellectuals; it seems to be more and more architects. You have somehow anticipated a lot of that because you very early on took a very public position, and carried it across the fields also, not just as an architect but as a curator and so on. I was wondering if that idea of a public intellectual was something which mattered.
Well it does matter because the intellectual has the ethical obligation of commenting on the society and helping people understand a number of seemingly disorganized events and to give them a certain structure to better comprehend it and, if possible, even to teach them to understand it themselves, so the intellectual doesn't give them just a fish but teaches them how to fish. That's a different story because very few intellectuals want to do that. If you think about someone like Sartre who comes from the great French tradition which invented the soufflé, am I right? you can take a

little egg and make it into a big book. One idea can be beaten in so many different ways and heated up. Provided that when it is taken out of the oven you have preheated the room — would you allow me to suggest I am talking of something like a publisher's advertising campaign, the soufflé can have a long life. Many intellectuals are masters of that and the public ones even more. Do I have any trust in architects being intellectuals? No. To me they have always been quite unintellectual including me. I am not an intellectual, certainly not. As for those architects who fancy themselves intellectuals, I find them, usually, they do not trust themselves because very seldom do they have an analytical formation that allows them to present things with a certain amount of rigor. I had a case (and I will not mention the name, just let you guess it) that exemplifies what I am saying. I went once to listen to a lecture of this person at MoMA and while I was listening to the lecture I was saying, "Am I so tired that this is for me a case of déjà vu?" I was holding in my bag George Steiner's *In Bluebeard's Castle*. I took it out and as he was reading I found the page that he was reading. Wherever Steiner wrote *literature* our personage was saying *architecture*. That was his lecture! Aware he was a second-rate intellectual he at least paraphrased something good with a tiny bit of falsification. But it was creativity and falsification. Of course the situation has to do with the fact that architecture having become an academic denizen needs to parade at least as an aspect of the sciences. Not having been able to develop a body of discourse, and turned it into a transmissible body of experience, architecture schools are, in most cases, glorified mini professional offices with the latest published competition being touted as the teaching program. That is the reason why I was so interested in helping to create the Institute for Architecture and Urban Studies. But my Institute was to be a station between the university and the practice of the profession.

Can you tell me about that because I am confused about the genealogy of that whole idea of the Institute and it is something which is related to the exhibition, because exhibitions are somehow schools. Yesterday I was speaking to Bruno Latour, who has now curated two important shows. I was involved with the first one, with *Iconoclash*, but he has now done one on his own, this show *The Parliament of Things*, and obviously the great previous example would be Lyotard with *Les immatériaux*, so actually philosophers would suddenly use the exhibition and make it into a school. Before we talk about your exhibition at MoMA, what I presume can be presented as an example of the exhibition as a school, maybe it would be interesting to talk about that whole idea of the school in more narratives because there seem to be different —
School?

Well, I mean the *Universitas Project*. There seem to be different moments of your involvement in that topic because it appears in your text, actually, which is here also, if I can manage to find it. Yes, here it is, the *Universitas*. In all your early interviews you always talk about the lectures you organized there: you invited Lefebvre and many other luminaries of that time. Now, at the very beginning of this conversation before we recorded, you mentioned that you are working in relation to that MoMA discussion on a new university, so I was wondering if you could fill me in a little bit on how that whole idea of knowledge production evolved. There have been numerous schools and you have always been founding them, if I understand well, rather than just assuming a professorship in an existing structure.
Yes, I have to say I have an immense admiration for figures like Manco Cápac, who was the legendary founder of the Inca Empire, and other such figures that come, create an institution, and one good day they take their boat, paddle down the lake, and disappear. My feeling is always that I like that notion. I make it and then it should survive when I disappear. Basically, my concern was, and the reason, as I told you, why I went to MoMA and I didn't stay as a professor at Princeton, and all that nonsense that is flattering when you are twenty-five years old, was because I wanted to get involved with the reformulation of the notion of university. Just let me clarify. We have had the philosophic academies with the Greeks, which in some way developed methods for understanding the cosmos and men by means of myths and legends. We have had the scientific universities that gained mastery over nature and proposed normative standards to be approximated which, with our mastery of the scientific method, we could perhaps achieve. But I feel that we have begun to populate the given nature, *la nature donnée,* with a man-made nature. Nowadays, if we see a tree it is there because somebody placed it or because someone left it there. We have created an artificial nature. By artificial I mean made with *arte e oficio*. And that artificial nature, or man-made nature, has to coexist with the given nature through a constantly rewritten pact of reconciliation. We know perfectly well that reconciliation between husband and wife usually lasts a few hours, a few days, a few weeks, but it crumbles after a while and you have to remake it every day. So my concern was, "How do we reformulate a notion of the *universitas*?" And that was, of course, the basis of the fable entitled *The Univercity,* presented as a city. But it was presented also as a word play on the etymology of *Universitas,* which stands for the whole, the totality. That's why I went to MoMA; it was because I wanted to create, physically, a new type of university, a *universitas,* using the museum opportunistically as an institutional support for me to be able to do it. And I got it. We went very far with that. It started to happen in 1970. I wrote something called "the black book," which was entitled "Universitas: Institutions for a Post-technological Society" and printed in only a hundred copies. And then I invited a number of remarkable individuals to criticize the ideas therein contained — but to arrive to that book, of course, I also had the help of a number of remarkable people, like Ronald Dworkin, who I consider one of the best legal minds in the nation, especially on the subject of the right to civil resistance, which touches on the notion of passive resistance.

Let's talk about your long-standing concern with the practice of passive resistance as a peaceful method people can use to stand up against injustice.
The notion of passive resistance has concerned me since I was fifteen and I read admiringly about Ghandi's life. The first practical enactment of the possibilities of such a powerful tool as passive resistance offers came about, by chance, when I was fifteen years old. I had a friend who was older than me by twenty years who had bought in Switzerland a machine to make industrial type carpets automatically and was fortunate to have sold the whole production of the

forthcoming five years to Peugeot Argentina. Obviously, he didn't have to concern himself with his business anymore because this machine was weaving the carpets by itself, so he took my advice he should just advertise his company's name, *Alfing* following a scheme I developed. One good day, at 9:00 pm all four TV channels of Buenos Aires went off, or so it seemed. People started wondering, very concerned; they didn't know what happened to their beloved TVs for which they had spent so much money. About thirty seconds later there was still no image but a voice came through saying, "This is a minute of silence offered by Alfing Company to return to the good habit of talking at dinnertime." The next day, at exactly the same timethe same thing happened again on the same four channels. Of course, people were aware by then as to what was behind the silence this time, and they didn't worry, at least for long. Ten seconds later, they could still see nothing, but a voice stated, "Please raise your hand; yes, yes, please raise your hand." A few seconds went by and the same voice came back to say: "In this moment twenty-five million Argentines are NOT raising their hands." The third day the program was canceled by the military junta because they said this program was a subtle device for training people in the techniques of passive resistance. I hadn't realized I was training anyone for passive resistance but they gave me this marvelous idea.

We also had Robert Nozick, who couldn't stand further away from me because he was a libertarian. Poor Robert died almost a year ago. He was opposed to the notion of institutions but I wanted him very much as a counterweight. I had a number of people as advisors in the formulation of that black book. They are all listed in the book recently published by MoMA. They read the previous drafts of the so-called Black Book, criticized it, and once I thought it was properly formulated we sent it to the invited critics. So I sent the Black Book to a number of people whom I wanted to read it and criticize it. The method used was that the invited critics would write a critique, and then their own critique would be criticized in turn by another designated invited critic. I told them who the critic of their criticism would be. Then we met on a weekend in a closed session at MoMA, no public except Robert Silvers from the *New York Review of Books,* and people from the Ford Foundation and the Rockefeller Foundation, who were genuinely very concerned with problems of education. So those are the people who really interested me. So that happened in April or May of 1971—don't quote me on the date—but it was about that time.

At the same time I was also working on another aspect of the plan. I had come to realize I wanted a physical institution that would have a new town as its laboratory. So we went so far as to try to create a new town. The idea was it would happen near Jamestown, about seventy miles (a hundred kilometers) north of New York on the Hudson. Nelson Rockefeller was still the governor, but he was also the protector of the State University of New York, which he wanted to become as good as Berkeley, California, a sister university. The goal was that this new town would be owned by the university. To that purpose I unearthed and found still legally valid the old Land Grant Act of the United States of eighteen hundred and something under which many agricultural colleges were created in different states. These colleges were in reality no more than disguised scientific universities, but you couldn't tell the senate of that state that you wanted to create a scientific university; you had to tell them you were going to make an agricultural college that would teach the farmers how to cultivate and how to get better results for their production. To my delight I found that the Land Grant Act, whereby the new university would have been given seventy thousand acres of land (a hectare being two and a half acres), was still alive. So instead of having a Land Grant Act we had the same legislation utilized as the basis for having an Urban Grant Act and the university was given an Urban Grant Act using those seventy thousand acres as their capital. The notion was that the university would issue bonds backed by the land as collateral, but we would never sell the land. That was one of the major requirements because I believe many of the deep problems we have in urban planning come from the fact of piecemeal ownership of the land.

So in this sense it was a bit like Buckminster Fuller: the land would be more a service than a matter of ownership.
Yes. Thank you for quoting him and not somebody more to the left! Yes. I have always felt that it is impossible to do urbanism in a serious way and on a proper scale by just dealing with and trying to reconcile the interests of many small parcels' owners. Norman Mailer was very right when he was running his campaign to become the governor of New York State and he said he would invade New Jersey and parts of Connecticut and Pennsylvania because, indeed, it is impossible to solve New York State's problems of water and air pollution, if you don't have a regionally integrated program. And the notion of fifty states or fifty-one states is something we inherited from an agricultural past not served by today's communication possibilities. Such a set doesn't any longer make sense. So one of the people to whom I sent the *Universitas* project Black Book for review was Rexford Guy Tugwell, who had been one of the brain-trustees for Roosevelt. He was still alive. He was the man who proposed that the American Constitution had to be rewritten (that he did in the late thirties) to reflect the fact that America was becoming a system of regions, no longer an economical system of states and that it was this system of states that was inhibiting the capacity for introducing certain necessary changes on a large scale. OK. So coming back to the subject of the new university, Nelson who, of course, was a prince, managed to get the high-speed train—going from New York to Boston, to stop at Jamestown. That meant that you could be a faculty member at Jamestown and in forty minutes you would have been in New York, which was very important.

So basically it was a school of urbanism, one could say.
No, it had many other programs: it was to have programs in preventive health care, public transportation, housing, schooling, etc. I was interested in transportation programs and in creating a notion of land ownership based on the concept of leasehold, which is a great tradition in Britain. But I wanted very much to get involved in the fact when a baby is born we can take care of him and maintain life long records of him and , thusly, we can, perhaps, predict certain problems that may occur to him. That was thirty-five years ago, when genetic studies were a figment. So, for me, just being a designer of the container without taking care of the contents, was for me a fallacious notion; I wanted to have

the whole. So we were really very interested, among many other subjects, in problems of health care.

Were private developers involved?
No, no, no. Developers, by the nature of their investment, have to think in short term for returns. It is for the state, it is for society, to make the socially necessary mid- and long-term investments. I wouldn't get developers because, naturally, they are not angels; they are card-carrying members of the human condition and they want their reward in this world, not in the next.

You also, I suppose, had to deal with lawyers, didn't you? One of the things that I have always been wondering in terms of your work, because I didn't find much in the books, in terms of your whole activity with patents there must be some form of involvement with the law. In terms of transdisciplinarity it is an interesting question, from architecture design to the lawyer. Have you studied it yourself or did you work with lawyers for that university project and for the rest?
I am tickled you asked this question. I never studied law, nor did I ever read law books, but having to obtain patents I had to understand how a lawyer thinks and would structure a claim. A claim is a description of a reality yet to become and it is a codification of said reality so that it can be compared with other codes of the same sort. So every time I had to present an idea to my patent lawyer, an exceptional person who, regrettably for me, has retired, I had to present my idea in a way that would help him organize and present for patent examination the yet-to-exist reality that is an invention. The invention has to be introduced into and fit congruently with other realities which already exist. The patent lawyer's task is to show that it doesn't conflict with them or steal from them if they have proprietary rights. That mode of thinking has always interested me immensely. I always have problems, for example, when I have to deal with situations under Roman law because in Common law everything has to be formulated and stated, whereas in Roman law you have jurisprudence, you don't have to say everything. But, for me, Common law, where everything has to be described and all possibilities have to be predicted, and sometimes forestalled, is a superior system of thought for claiming a reality yet to become.

As in the US.
Yes, and in England, of course. It is a very interesting way in which you have to focus, as with the design. You have to be sure you don't create situations where you have secondary or tertiary results with an unexpected and unwished and unwanted notion. In case these occur, as they will inevitably happen, you have to design your structure, whether it's a product or an idea, in such a way that you can remove the problem that emerges unexpectedly with the minimal amount of damage to the whole of the structure. So in a curious way it is almost like system thinking, which I have always been very fascinated by. The first thing I ever wrote — and the last I did on a theoretical level — had to do with method-idolatry and it was called "Notes Toward the Formulation of a Design Discourse." It was published by Yale in 1968. I think it was issue number 12 of *Perspecta* magazine. Several respected people involved with design methodology have told me that this article had a great influence on them. I was very willing to believe them. [...] Allow me to come back to the *Universitas*.

We have philosophic methods of analysis, we have empirical methods of observing reality and mastering it, we have methods to formulate norms, which is probably my connection with law. What I think we don't have yet is methods of thought leading to invention because this involves a synthetic act, a jump into nothing; we cannot deduce it. We have to develop a methodology whereby once that jump is taken we can interpret it; we can in some way get into the kernel of what is wrapped in the image we made.

If one looks at inventors it sometimes takes quite a long time and effort to come to the first inventions and then suddenly there are hundreds of inventions. Does this have to do with this?
Yes it has to do with that, of course. For example, when I invented the Vertebra, which was the first automatic ergonomic office chair, the very first one, in 1974 and presented it publicly on June 23, 1976, there wasn't another one. The manufacturer who licensed it told me, straightforwardly, if I had not engineered the chair with the help of a colleague, if I hadn't brought him everything done, with the tools, with the dyes, with the machinery, he would not have taken license. Even then when I presented it he was very frightened that when he presents a chair that moves people will think it is badly assembled. I remember telling him, "Let me do another chair of the same spirit, of course different mechanisms, for another manufacturer because at present you have a hundred percent of the market but the market is minuscule because the culture hasn't yet accepted the idea of an ergonomic chair. He refused. Others were later inspired by that idea and made other chairs, similar in behavior but based on different mechanical principles from mine. When this happened my licensee came to have 10 to 12 percent of this emerging market but it had by then become immense. So now I have many children; it has generated a whole industry. And that is, of course, the notion of an invention. Some of them are so essential that they can only allow variations; some of them are just improvements on something existing.

Like the lamps you designed and Erco Leuchten produces, I suppose.
The goal was, basically, to have a lighting instrument so designed that you can actually predict how the lamp will behave and illuminate before you put it up on the track. It was designed for museum use. Having been a curator, I have been on enough ladders trying to aim a light so that I would have been delighted if I could have a product I could preset on the ground, and then have somebody else put it up on the track. I wanted also something so designed that it could never end up aiming up to the ceiling because if a lamp aims up to the ceiling you might have a fire. So the lamp's body is a cylinder, elliptical in section, which has been sectioned at forty-five degrees generating a perfect circle, which I had graduated so that it can be preaimed at any of the angles inscribed in the circle.

Around the same time of the university idea in the early seventies you also wrote that very fundamental text of the fable *Designer/Producer,* I think it was in 1971. I am extremely interested in one very important aspect of your work which resonates in a lot of artists' practice, which is that right now a lot of artists, if you think about Rirkrit Tiravanija, Philippe

Parenno, Pierre Huyghe, they are very interested in this idea of production of reality. They say exhibitions are possibilities to produce reality and then the next step is to produce reality. I was fascinated by your ability to actually produce reality. My question about the law had to do with that and you beautifully explained how the law is part of that production of reality. I was also wondering how you would define that production of reality and when that started within your work. We can then return later a bit to the chronology: we were with the early beginnings in Argentina and then we talked later about MoMA. It is very often in the books that then you stop being a curator and more and more your exhibitions produce reality; if we think about the *Taxi* show, much more than the Barragán show, it kind of produced reality and then from the *Taxi* show it was the next step to be an industrial designer. But I was never really sure if that was the chronology or if maybe in the first work you already have everything. I was wondering if already in Argentina that idea was there in your early work.

I was a child with a fixed idea, an idée fixe. I had only one idea: I would be an architect. I was eleven. I was fixated! Nothing else would do for me! Of course, when I was asked what I would like to be when I grew up I replied I would like to be an architect and a gigolo. I think I failed at both. Most regrettably I missed it as a gigolo, which really would have been great! That would have given me great pleasure. But it wasn't such an esoteric combination of goals for one needs to maintain oneself; but anyway, to come back to the story, when I was fifteen, I designed my first house. The client was a couple of teachers who lived across the street from me. One day I woke up with an idea perfectly designed in elevation, in section. I never design in plan. I always design in section because that is the phenomenological way one perceives things. It was a house with a blind facade that was just a wall, with a dissimulated way for entering the wall and a way of "walking the wall." You would have to walk steps cantilevered on the facade to arrive at a loggia-cum-balcony. It could only be reached from the facade; you couldn't reach it from inside the house. Anyway, it was a very cubist building before I even knew about Le Corbusier or Cubism. I am always promising myself I will build a model of that house. Anyway, architecture was my obsession and so I stated in my application to Princeton. When I met the officer in charge of admissions he said, "I finally meet you. You are the first person who ever made an application to Princeton stating: 'I wake up as an architect; I go to bed as an architect.'" Of course, I was not only a boy with an idée fixe, I was also a very pretentious little boy! I went to a high school which was a technical school, very much copied from the Swiss Technique schools who used to prepare *capo cantiere* (foremen).

Was there a Max Bill spirit around?
No, the Max Bill spirit that was to be was already anticipatedly lodged in Tom Tomás as Maldonado; he was Max Bill's ambassador before Bill existed as such, his representative on Earth describing a domain that existed somewhere far away. I do think that Tomás was a far better painter than Max Bill; he is certainly a far greater theoretician than Max Bill. You must understand that Tomás worked in 1955 creating concrete art. He needed a foreign father, a very Argentine weakness. He was already fully grown up as a painter when he declared Max Bill a father! He adopted a father so his phenomenal talent would not frighten anybody as coming from nowhere. As a matter of fact in 1958 or thereabout there was an exhibition in Germany entitled *Acht Argentinische Abstrakten* showing the pioneering work of a group of Argentine concrete painters and sculptors that anticipated many of later European concrete art movements. Among these artists were Hlito, a very lyrical painter, and a marvelous sculptor called Ennio Iommi. Their work forms part of the collection Ignacio Pirovano donated to the Museum of Modern Art of the city of Buenos Aires (MAMBA). My burning desire that said collection, hidden in the museum's basement for lack of presentation space, be shown is the reason why I donated the architectural project for the MAMBA new building. All those artists were people I admired greatly when I was a kid; I must have been fifteen. They were great artists and their work still, today, is very moving. It moves me, perhaps, for nostalgic reasons, but when I ask other people who know their work, they all assure me they were really very genuine new artists. In Buenos Aires we were always extremely attentive to what happened in Europe and America. In 1955, Kenneth Kemble, the English name of somebody born in Argentina, probably from Welsh parents, was doing pop art avant Andy Warhol. His work was exhibited in the then Museum of Modern Art directed by Rafael Squirru.

I made my first exhibition when I was fifteen not as an exhibition curator, but as an installer. It was an exhibition I completely installed—I just installed it—when I was fifteen, at the Museum of Fine Arts. It was glassware from Finland, things designed by Timo Sarpaneva, Kaj Franck, and Tapio Wirkkala, among others. I did the installation. I fancied myself more than an apprentice. I know now I have remained an apprentice.

But to come back to the other question, where did the industrial design story start? I became a curator by chance; I was never curator material, I was impresario material. OK, I was also concerned that I wasn't paying too much attention to MoMA's collection. There were certain lagoons in the collection and I was very happy that I had another colleague, Ludwig Glaeser, who was much better trained academically, who could in some way carry on with these scholarly tasks because I was much more interested in creating an event, in making things come about. I used the museum's support to help bring about something which didn't exist and could not exist unless we provided the institutional support. You mentioned the *Taxi* show. That was one absolutely one hundred percent case where I used the museum's prestige to bring about things that did not exist. Is that a good formula for all curators? No. I am very distressed when a design curator produces an exhibition just of different materials showing its potential because it's like if my colleague in the painting and sculpture department were to do a show of new types of acrylics, new types of colors, I think that a painting and sculpture curator has to wait a little bit for the artist to produce it. He can fancy himself an artist, like Bonito Oliva does; by the way he assembles the artists and their work: the curator as bricoleur. But in my case, being a creator of design, I could in some way be the spark that provokes, providing the humus from which things are to sprout. I felt that such was the role I should have: not only collecting things that were made but helping things to happen, not so much to increase the collection but really to increase the level of awareness and being able to allow designers to operate in a domain that normal commercial life would not allow them. So that inspired me to write that little fable you mentioned, *Designer / Producer,* and as I was writing I realized that I had already starting thinking about doing industrial designer as both creator and producer. I resigned from the museum three times, before they accepted my last one, before *Vertebra* was presented publicly on June 23, 1976, because I firmly believe, to this day, that a museum curator should not be a gallerist or practice as an architect or a designer. I find it a conflict of interest by which I do not abide. Hopefully this interview will appear after my exhibition at MoMA. When I was a MoMA curator I didn't even accept to be invited for a cup of coffee. As a curator I was a judge and as such I felt I should be inaccessible to any entreaties. My greatest pleasure as a curator was to discover seeds that needed watering, as I did with people like Rodolfo Machado, Tod Williams, the Morphosis boys, and Stephen Hall, who were completely unknown kids. When I left the museum I became the president of the Architectural League and I created two series of programs. One of these was "Emerging Voices" for architects thirty to forty, and then another for "baby" architects (obviously using a different name) for kids that were still working in an office under someone. I had an immense pleasure when discovering someone had talent and give him / her the proper conditions so they would bloom.

To come back to the subject of the designer as impresario or producer, I have always felt very strongly that to separate the industrial design profession from the engineering task was an eminently artificial division of intellectual labor. The Bauhaus's mirage was that when someone finished his training as an industrial designer there would be waiting for him / her an enlightened industrialist and the two together would do marvelous things. Regrettably, the phenomenon represented by the association of Peter Behrens, the architect cum industrial designer, and Walther Rathenau, the farsighted industrialist of AEG, wasn't easy to replicate.

What can be done?
The designer has to become again an industrialist. The Bauhaus hindered a process that had started successfully with people like Thonet, Singer, Ford, and many others. They were inventors-producers. There was no way that their inventions would have been produced if they did not also have a talent for engineering and production, as well as for getting protection. Thonet got the protection of Prince Metternich, who gave him patents in exchange for shares in his company. Oh, Metternich he did get for himself and family a longer-lasting empire than the one he defeated. He didn't defeat it in battle but he was the one who gave the lie to the notion that what the generals win in the battlefields the diplomats lose at the conference tables. With him it was the others who lost! Anyway, to come back to the notion of the designer engineering his own inventions, I have always felt that the designer-producer was a figure that had been stopped by the artificial separation of intellectual work the Bauhaus established. That I wouldn't accept.

A segregation, basically.
Exactly. And I felt that the notion the designer that makes it beautiful and the engineer that makes it healthy is an immense fallacy. I have been working for twenty-seven years as the chief design consultant to a company that makes diesel engines for trucks and trains, Cummins Engine Corporation. Naturally, they called me to make them pretty; something like: "What color, Mr. Ambasz? Shall we put the wires here or there?

The company hired me, probably, to choose the colors and to make the wires run straight but, instead, I made a promise to the owner. Before I tell what my promise was I should tell you he was one of the most enlightened men America has produced in the field of manufacturers, his name was Irwin Miller. He died, regrettably, a few years ago. He was an immensely illuminated man; his feeling was that you do culture in the factory, not after 5:00 pm, and that the products his company made had to withstand being examined a hundred years later as valid cultural statements by anthropologists. So I said to him, "Look. I am not going to design anything. I am going to come here for three days every two months to teach your engineers to recover the trust in themselves as designers. The engineer is a designer and vice versa. In short they are one and the same. I am not here to help him do the lonely jump that inventing demands; he has to jump alone. I am here to provide him with a system of questioning himself so he can, once he had an idea or an image of an idea, develop it, further it or, perhaps eliminate it if is not good." And for twenty-seven years the engineers of Cummins have designed every product for which the company won many industrial design awards. But, as we know, common lore demands that a promise be broken, so I did design one thing myself: the oil cover of one of the engines. That's the only thing I designed because I had promised him, "I will not hold a pencil in my hand. I will teach design to your engineers; they will be the designers. In the end they will do their own designs, invent forms that best serve the function the component in question is to perform; forms that are simple to manufacture, easy to assemble, easy to service, and by so doing they will be inventors; a more comprehensive definition of their task than engineer or industrial designer." Because, again, I have insisted all my life, and I insist here, designing is an act of the imagination; designers / engineers should be inventors, they are not academic titleholders. Invention is a state of the spirit, not a state of diploma.

Of course the system in America, and in many other countries, punishes whoever makes a mistake and therefore the engineers working at Cummins were very frightened. That was my role. I told them, "I am here to cover up for you. When they ask you: 'Why did you do that?' you say, Emilio told me." "I am here for you as a phantom because if I am fired it doesn't matter to me." And people think that I have power, which I don't really have, but which I detent mightily by never exercising it.

That is very interesting because that actually connects to something you said previously. We are jumping, but that's no problem; it's nice that it's not linear. The issue of instructions is super interesting. It is something I wanted to ask you before; it has come up now two or three times. I was very

interested in this whole idea: John Cage spoke about the open score; Yoko Ono did a book only of instructions; the whole Fluxus movement, influenced also by Duchamp and by Cage, had that idea of a very open score. I have the impression that in your work the whole idea of open instructions or open score seems to play some form of role. You said you are not really doing the drawings and at the same time you are basically saying, "Here, that goes even further." What you were just describing has to do with not even giving instructions but more setting up a structure.

The key word is Method. It comes from the Greek word *metodos.* It means "the way," *el camino,* and if I may answer your question in an indirect manner, allow me to say that the first year I taught at Princeton, I was given the freshmen class. I myself studied at Princeton, well "studied" is what I claim. Ada Louise Huxtable said I should have stayed there longer rather than complete the 4 years of college in one semester. The undergraduates at Princeton are immensely bright, intelligent; they are also immensely uncultured, but, as I said they are intelligent and inquisitive. Graduate school is another story. Let's not start on that! And what I taught them was not Design 101, but Methods 101, using design as a way to solve not only design problems which on first examination may seem to be composed of elements which are dissonant and dissimilar from each other, by organizing these into a congruent structure. You know that you don't go to college in America to become an architect; you go to college to get a general education, which in Europe is supposed to have been acquired in the *Gymnasium,* but even that is no longer the case in Europe. Certainly the high schools in Europe produced people who are much more cultured than the high school kids I see in American universities. But I had students who thought they wanted to be doctors or lawyers but they were not sure if they wanted to be something else. Basically, they didn't know what they wanted to become. So they came to take a course in architecture because they thought they probably wouldn't have to read too many books and they were told by other classmates who took Architecture 101 with another professor, that it was just an easy thing and it would give them time to keep on running after girls. So I taught them Methods and the vehicle I used to teach them was using graphic design, product design, and architectural exercises employing very short problems lasting one week. At the end of every week, when they thought they had mastered the problem assigned, I reviewed and criticized their proposed solutions. And the next week I gave them the same problem again and with what they thought they learnt during the week before they thought they solved it this time and I proved to them again they didn't. But I was never doing it as an act of virtuosity or acrobatics on my part. I was always trying to teach them the methodological why of their failure, and how they could have thought to organize it. My immense pleasure is now, when I meet some of them by chance, already doctors or lawyers, they always tell me that my course had a great influence in helping them to organize thought. And that brings us back to my curiosity about how lawyers think and elements of that sort. This method of teaching was appreciated by my students who voted my course the best in an evaluation test the university conducted, and I was made Philip Freneau Preceptor at twenty-five. I resigned from Princeton a few weeks later. I resigned for two reasons; first, because I wanted to go to the Museum of Modern Art in New York to put in action my ideas about the *Universitas Project* I already mentioned, guessing the Museum would provide a platform making that task more possible. The second reason was I couldn't face those kids any longer knowing that the manner in which they were being taught in the school was so inappropriate. It lacked so many things; it lacked the understanding of the role invention can play, the teaching of design as an aspect of the creation of a man-made environment of which architecture is just one component. A more comprehensive approach to teaching would have required a far more interdisciplinary approach. That couldn't be done because universities are not set up for the integration of disciplines; they are set up by departments. Their premise is that a perfect world has been created by a supreme entity and therefore we just have to understand it by segmenting it, analyzing it, and such process will allow us to understand the whole. That Cartesian approach to the constantly changing configuration of the universe is the basic fallacy on which all universities exist nowadays. Ah, Baruch Spinoza, why have you been neglected?

Julia Kristeva recently said in an interview, there is a fear of making a pool of knowledge and that fear is paralleled in the universities. In relation to Argentina, I have conducted long interviews with Gyula Kosice. He is an amazing man, and what is interesting is that there was architecture, graphic design, there was writing, there was theory, and he was always talking about his "laboratory years" in Argentina. The thing I was curious to know was if you think that had something to do with your laboratory years of being brought up in Argentina, before leaving, and that whole climate. Because also that link to literature was there in Buenos Aires, super strongly. There was a quite interesting architectural context, as far as I understand also from Kosice. He told me there were a lot of visionary utopias around. I would be really curious to know how that climate Kosice describes has been interesting and what your points of connection were. In your interview "Ambasz and Emilio" with Michael Sorkin you single out Amancio Williams as one crucial influence. It's a big question.

Argentinians—as I said in that interview—we are the greatest bankers of information on the planet; we import information, and we reelaborate it. If somebody sneezes in Paris, (OK, no longer does anyone sneeze meaningfully in Paris, granted), but let's say now somebody sneezes in some intellectually creative center, the next week you have a symposium around that subject in Buenos Aires. And I mean it seriously. [...] There is immense attention to what is being done outside the country. Our tragedy is that we know what the historical moment could provide us with, but since we are economically underpowered, rather than create groups to overcome such situation, everyone does on his own everything possible to inhibit anyone else from doing anything, least their success signify his life is a failure.

We were talking before about Tomás Maldonado and the Argentine constructivists and all those other Argentine artists who were so brilliant. Gyula Kosice migrated to Argentina— in a certain way he appeared as "the Hungarian inventor" as if he had been manufactured by a Hollywood office of central casting. He was a man of immense imagination and I think that the Argentine context didn't give him anything. He already came programmed from the "factory" to be that way. Argentina did provide him a situation where he could sit in the provinces and have a dream of the metropolis. That is the way the metropolis nourishes itself; it nurtures itself really from the dreams the provinces have of what a metropolis might become. If you look at the exhibition I mentioned before that took place in Europe in the 50s, the eight Argentine concrete artists, you will agree it was an absolutely extraordinary show in relation to other things that were happening, or rather not happening at that time in Europe. Presently, the economic misery that Argentina underwent a very few years ago was absolutely heartbreaking. But the extraordinary thing is that during this most difficult period there took place in Buenos Aires eighteen hundred exhibitions of art, and I don't mean just galleries putting up shows but also theaters and cinemas, as well as every facet of artistic endeavor; it was an immense creative explosion notwithstanding the fact so many people could survive only because they collected garbage and resold it. While the economic situation was absolutely heartbreaking, what was going on there was exhilarating. As I said in *Anthology for a Spatial Buenos Aires,* when living in Argentina you have the feeling of sitting on an immense plateau which starts in the Andes and ends up at the sea landing very far away, maybe in Europe.

In my case, there was a balcony on my parent's apartment facing onto a tree's branches. At night, when I walked out onto this balcony I could see the dome over Argentina the stars defined and this gave me a feeling of, "How lonely we are in the universe." I think some people of Buenos Aires have that feeling, that we are immensely far away from anything, and that to have something of our own we must create it. That type of invention of a reality of our own requires it acquire material existence. In that regard, Argentina's great tragedy is the absence of this materiality made by us. On one side we are very aware of what the historical moment offers us in Europe, in America, and in Asia. We know fully well what is going on, what we could have done, and at the same time we do not have the economic means to put it into practice. That creates an immense amount of tension, which in healthy societies would have been resolved by the people getting together, making groups and trying to change things or trying to do things. In Argentina, in my opinion of course, it resolves itself in everybody retiring within himself but making sure that nobody can do anything, because if somebody does something and brings it about that would be a proof that everybody else's life has been a failure. That is our tragedy: the incapacity of existing as a group, as a society. We exist in Argentina, and it is a cliché, but it is true, as we say in Spanish, as *aves de paso,* birds that come, eat, and then fly away. We all come actually from Europe and other places but no other country has decided like Argentina that it will not become a culture except by those excessive nationalists, who feeling the immense need for that and not having the capacity to invent that culture, go into modes of extreme nationalism and assume an intemperate nationalistic I.

What I am going to tell you is unrelated to what I just said. As I already told you, from the time I was eleven, I wanted to be an architect. I had a friend who was working for Amancio Williams and he talked to Amancio about me. I was that time already fifteen, and still in high school but this not withstanding Amancio accepted me to work with him. You must understand I was in the fourth year of high school, something like that, and he accepted me. So I changed my life as a student; I started going to high school at night so that I could work with him. Why did I go with him? I went with him because I thought he was a great poet and I think that Mies van der Rohe and Le Corbusier thought also of him in that way. The letters from them prove it. These two men were Europeans, but for them Amancio was a man who could perhaps create in the new continent the utopian models that they had thought of. They fancied that their models would be interpreted and erected in the untarnished, unrestrained, unbound land of Argentina. I feel strongly that in some of his projects this did happen. It was like the elaboration of modernism taken to its second generation. You remember that Bach's children used to call him "the old man with a wig," and he, of course, belonged to a late generation of wigged baroque musicians, but what magic synthesis he created, what quality! There was no historical change there, true; rococo was already ruling the music schools, starting with that of Mannheim, but was not Bach's an immense artistic achievement? Who cares if Guercino, greatly influenced in his youth by Caravaggio, painted in said style forty years later? His are still moving paintings. Anyway, coming back to Amancio Williams, I wanted to work with him because he was a great poet and he gave me great chances. The first thing I did for him was to go to Uruguay to make the arrangements to build a building for a great patron of art and architecture, Guido Di Tella, the man who sponsored the creation of the immensely influential Instituto Di Tella, a great nest of artistic activity. His father made a great industrial fortune and he used this fortune, really in very generous ways.

That's what I meant before about the laboratory. That exemplifies it.
Well that was, of course, one of the real physical laboratories. But Kosice was working before the laboratory was made and the laboratory comes about when you have the chemist and he was already making chemistry in his kitchen. He had sculptures made of water and he was made fun of by many artists. Artists are most ungenerous to each other. But, then, so are architects.

The notion of the water is interesting. If one looks at your work, water is a very persistent element. One might almost say water is a medium and I was wondering if you could talk a little bit about the feeling Argentines have for water.
You must remember that in Buenos Aires we have the River Plate, and this brown river is as wide as an ocean. At sunset you get extraordinary greenish-brown colored clouds at sunset, a reflection, I like to think, of the pampas rather than the river. You must remember when Ray Bradbury talks in *The Martian Chronicles* about Mars's brown clouds.

The clouds are colored.
Yes, the clouds are colored at sunset and they are not the beautiful pink, orangey-lilac colors that you can see in Venice at sunset, they are green, and they are brown; they are reflections of the pampas. The city may be big but the pampas are bigger so the reflection comes from there. I do not remember who, if not nobody, who said that Buenos Aires

lies between a sea of pampas, the green sea, and a sea of brown, which is the river. I am in love with water because I find in it greater depth than what its surface reflections belies. I believe strongly that one can evoke the presence of architecture using noncanonical architectural elements. I believe you can evoke the presence of architecture even by using bales of hay. [...] You don't have to use the columns or any of the elements comprehended within the academic tradition of architecture to conjure its presence. Water is a living entity such as grass is, therefore I make water to cascade up, like in one project where I wanted to show water climbing steps. That was my way of inviting people to walk up the steps of the ACROS Building in Fukuoka. Regrettably, this feature was "improved" by the Nikken Sekkei architect in charge of the project. [Sighs] His forgotten name was on my lips but my unforgiving teeth got the best of him!

Anyway, to come back to the story, yes, water plays an important role in what I do because it doesn't have a shape of its own; that is to say, it does indeed have an immense power of its own, but the shape it adopts is the shape of the container you give it. To me water is important because it can be nebulized. I have used fog many times to evoke the presence of a building which isn't there, and its presence becomes very strong when the sun creates a rainbow. It cools you or warms you if you make those clouds of mist. I use fog and its indeterminate form maybe because I am a prisoner of my time and afraid of making definitive statements. I seek to make statements which are constantly being reformulated. I always say there are two ways to cast a shadow: one is as a tree and the other as a cloud. I think that I chose to be a cloud.

So it is less the water as a medium; one could say cloud is the medium.
I am seduced by water's capacity to change shape.

Forgive me, this is a nice interruption. I think I am a bad host. Would you care for more water, orange juice?

Coffee would be better.
You can have both. I do not have cookies for your coffee or tea, but I can quote you instead Rem Koolhaas's — perhaps a bittersweet cookie — and his concern on how the market economy influences (or perverts) architecture.

I find that the market society that he feels is governing architecture — and Rem is one of the outstanding examples of riding with the surf, if I am quoting him properly — that means the architect has given up what I consider his ethical obligation to make the waves. He wants to ride with waves that have been made by circumstances. Life is easier that way. You can ride high; of course you can also fall from a height. I have found that such is the position adopted by those architects who believe very much in providing new dresses for those who have a mode of existence that can be more or less described as "I consume, ergo sum." He is, to me, one of the greatest exponents of that group, brilliant as he is, an extremely intellectually competent man. Rem should have been the governor of The Netherlands, and the Low Countries would see gold again.

That is actually interesting because he is considering venturing into politics.
Philippe Starck also wants to enter politics. It is quite understandable. When people have made a certain statement, and they realize that what they propose has not happened because they do not have the power to make it happen, they seek power, forgetting they are the slaves of the system, or to put it in Rem's terms, at the mercy of the sea. Like many slaves they fancy themselves smarter than the owners — and many of them are so — and they want to play the owners' game. I wish to point out that this I have already observed many years ago when Tomás Maldonado left the Hochschule für Gestaltung in Ulm, where he was the director. He left the school; bought himself a Jaguar, and he and his wife, Maya, drove all the way from Ulm to Milan. I was sitting in the car's back (and I assure you this little aside story is pertinent to the question you posited) and I asked Tomás, "What are you going to be doing in Milan?" He replied, "Well, you see, designers have great possibilities to influence events and I am going to be the design director (or consultant, or whatever the word is), of La Rinascente, a chain of department stores. I have a feeling that in that way I will be able to influence design much better and to bring about many more things than just by remaining as a schoolteacher." Sitting in the back, I felt slightly dizzy — maybe Miss Maya was driving very fast over the Alps, or maybe the reason was another. Taking a deep breath I said, "Tomás, how are you going to avoid what happened to other people? You remember the man who cultivated his grapes in the morning and at night went up to his tower with a candle to write his *pensées morales*? And if Michel de Montaigne couldn't avoid making references to what good wine his vineyard was producing, how are you going to avoid contamination?"

Now, of course, designers work with material which is part of daily life and therefore they have to accept the market as it goes, but in my opinion architects and designers have an ethical obligation, which I feel at present is not being fulfilled by the fact that many architects are being immensely successful — I mean the type of success that shines in their bank accounts — by playing the game of riding the surf. I think, on the contrary, that architects have the moral obligation to propose alternative models for the future. If not, one runs the certain risk of perpetuating the present. That is an ethical stance I am not willing to give up. So if someone calls my stance intentions, let him call it intentions. For me they stand for an ethical position. Of course I know that ethics in the market is a fancy thought, but I strongly believe that it can be achieved.

The question which is perhaps the most complex of the questions is the one about modernism / postmodernism.
That is my twist! I have no idea what Rem meant when he said I was a perverse modernist and then a naïve postmodernist, if I understood the question.

Oh, he is very flattering, he is very generous! I don't think I have been any of those things. The so-called Casa de Retiro Espiritual in Spain stood outside modernism and postmodernism, anticipating the deconstructionists by a long way.

Was that in the seventies?
Yes, it was designed in 1975 and built in 2000. I put in a closed balcony which I got from a demolition and which were historical pieces, and maybe, *malgre moi-même,* in that way I anticipated postmodernism. If it is true then I can only declaim "Mea culpa, mea maxima culpa." Of course the house had angles which were straight and walls which were white and perhaps because of that it could be considered that in some way it was at the same time also modern. We know perfectly well that even if I were to bite off my chain, I would run around loose but with the collar still around my neck so I can't completely get out from my zeitgeist, as they used to say. So I think that I have washed away any of the so-called sins of perversity and naïveté by trying in some way to create an architecture that evokes the presence of architecture without using fragments from the memory of architecture, without using modernism squeezed like a lemon and now twisted a little bit or cut up in different pieces. I have tried to show a way in which you can evoke the presence of architecture without inflicting any damage on the land, integrating it and giving people the pleasure of the land. I have gone one step further than modernism when it promised decades ago to everybody that they would have the house in the garden. For them the goal was that the plot of land be divided into 60 percent for the garden, and 40 percent for the house. I think that was a good idea. It started with Ebenezer Howard, and it then grew on and on, until we got Hans Scharoun turning this idea into the Berlin construction code. I want much more than that. I want, not the house in the garden, I want the house and the garden. I don't want 60 percent garden, I want 100 percent garden, and I want 100 percent house. I think I have shown ways of achieving it, not only in the middle of the fields where it is certainly easy, but in the middle of the cities, like in the case of the building ACROS in Fukuoka, where I show that you can build the building the developer needs to realize his margin, and the investors need to recover their investment and still have 100 percent of the land the building's footprint covers turned into gardens accessible to the public at large. You can do it in a way such that you don't take away one square inch of land or a centimetre of land from the people who once had the land for themselves (in the case of Fukuoka it was part of their only downtown plaza).

So now to the question of you as a curator. I have always seen your entering MoMA as some form of infiltration. For instance I was talking to my friend, an artist, in the early nineties, and he said the museum is an extraordinary tool and I should work in a museum. He was pushing me to work in a museum because he said a museum is a great key to society so if you want to make something happen, a big museum is fantastic, it opens every door. I always had the feeling that this is how you used MoMA.
I said it, I declared it. Unquestionably, I went to the museum because, as I said I wanted, secretly, to do this *Universitas* project and I discovered to my surprise that I was not a terribly bad curator. Like I stayed at Princeton to teach because it was one way of getting a green card and I discovered doing it that I wasn't a bad teacher. The only thing is that when you do something and you feel you have done it as well as you could, you start perceiving that you are repeating yourself. Move on! Do something else! I have been a curator and if they asked me now to be a curator again I wouldn't do it. I don't want to curate my show at MoMA. I just said "invite me to the opening," if there is one. But I went to MoMA, yes, as I said before, dressed as a banker. I could introduce the changes I wanted; I could carry it as far as I wanted. Had I appeared dressed as a terrorist, throwing ox blood at a painting I would have been shipped out.

Through that back door you obviously had to do exhibitions because that was the job. I wondered how you dealt with this at the beginning. What were the first shows?
The first show I did was something that I was asked to do by Arthur Drexler, the department's director. It was a show on Eugène Hénard, a remarkable French urbanist, based on a book that Peter Wolf had written. Please spell "Peter Wolf" correctly because if you don't I will never live long enough to see him twice. So on the basis of Peter's book and the images that he collected, I was immensely happy to make this exhibition as my first one because Hénard invented the *rond point* at street crossings thusly eliminating the need for traffic lights, he introduced the idea of "concave bays" in the middle of the street so you can sit in a café; he invented a number of things that could be achieved very easily without the need to have a new Alexander the Great invading and changing the city. It was possible to improve the quality of life with his ideas in a relatively short amount of time and that fascinated me because it was a good proposal. Yes, he was riding the surf accepting the status quo but he was making substantial contributions for improving the urban situation.

Then I got the idea of doing the Italian design show, very naïvely thinking I would go to Italy and just collect objects.

***Italy: The New Domestic Landscape.* That was quite an early one, 1972. It was an exhibition of radical Italian design.**
Well, no. When I went to Italy I didn't think of it being radical, I didn't even know anything. I thought I would just go and get objects and show them. I went to Italy and there was an extraordinary phenomenon going on there. There were designers who were really working within the system; one could call them the *conformists.* There were designers within the system who were trying to propose changes: the *reformers.* And there were the ones who contested the system completely and I called them *contestators.* I realized that showing only isolated objects would not describe the cultural role that design played in the life of contemporary Italians because they saw design as a way of making social criticism and even social postulations. So I invited a number of designers to create environments of a certain size and scale in three dimensions so they could go beyond designing the lamp, go beyond designing a little chair, and design whole environments. I also invited the *contestators* to contest. I invited your friend Enzo Mari NOT to participate, so I would take away from him the pleasure of posing by stating he wasn't participating. Of course, the letter says, "Dear Enzo, knowing that you will not accept to participate, I am inviting you not to participate. Rest assured that your objects constitute the largest number of objects by one designer contained in the museum's collection: ashtrays, plates, decorative elements, all those things that really are not needed by society but are an immense adornment, which is what society needs also." He is a great designer. No question.

So, they were all represented. There were also people who distributed political pamphlets. We printed 360,000 pamphlets of the contestatory group Gruppo Strum which had the face of Agnelli on the cover. So of course the Museum's trustees were a bit alarmed because on top of Agnelli's face it said, "Capitalist."

360,000.
Yes. There were three different numbers, so 120,000 of each, paid for, naturally, by Fiat. I think The Great Ironist must laugh his head off all night long, but then we are the ones who continuously provide him with material. So, the Trustees demurred saying, "How can we have that on the cover? He is a friend of the museum." I said, "I have spoken to Mr. Agnelli and he assured me that a capitalist is what he is." He had no problem with that. Anyway, the point was that the Italian design show had an immense effect on design in general, certainly American designers. They were all convinced by their Bauhausian ancestors that deprivation was the way to good design and here they had the Italians making curves and using colors and enjoying texture and a number of things that hit them in the stomach but at the same time seduced them. So it had an immense effect. Of course, it had an immense effect on Italian industry as well. I am still waiting to be made *Cavaliere della Polenta* by those ingrates.

It was interesting. It all came about after that you went to Italy. You didn't have a pregiven opinion about this show and that is actually something really essential. I have always thought that the interesting shows happen if there isn't that kind of master plan but when it is open and then suddenly one gets excited about something. Most museums do this wrong methodology: they illustrate the pregiven idea and it only leads to a bland show. What you say proves, actually, my conviction that the great exhibitions are not a master plan, they kind of happen.
It happened in this case because the objects I collected were there, I didn't make them. I knew about them before I went. My contribution was to make a number of environments happen. There was the environment that, for example, you can now see in the Milano Biennale on Joe Colombo; there was also an extremely intelligent thing: a house made out of a container for people who suffer natural disasters designed by Zanuso and Sapper, the cleverly expandable one by Rosselli, the car that Bellini designed was in a certain way a forerunner of Renault's later *Espace*. I gave those designers the institutional protection they needed to make their environments happen. I also raised the funds for them. But raising funds is very easy if you can communicate to the donor the sense of an intellectual adventure, to invite them to see the shape of tomorrow morning, not next year, just tomorrow morning. And of course, there was also for them in that the vanity of being present as a sponsor in the museum. That no minor fact helped greatly to raise funds. The important thing to me was that we were able to bring about these environments. In more than one way they were quite an important influence on design in general.

So, I went to the museum because I was only interested in doing the *Universitas project*. I had to earn my keep and I became, of course, an historian, and I started collecting entities. For example, everybody thought that the *Cisitalia*, an Italian car designed by Pininfarina in the period 1946/48, was in the collection. It was not. It had been shown in 1951 at MoMA but it was not in the collection. So I went to the Board of Trustees and said, "We should have a car. This is an industrial design collection. I know we have little space but you can't have an industrial design collection without a car." So we found one, outside Torino, owned by a farmer who was using it to keep his chickens inside.

A ready-made.
Not in good condition. So I called up Pininfarina's son, Sergio, and his bodyworks reconditioned it and gave it to us. So I started covering empty spots in the collection; examples of Gaudí's work were missing, we were missing things by many other designers. When I left, of course, I had a list of things that were still missing. After leaving the museum I bought myself some of these objects to donate to the museum. Mainly Horta; they didn't have anything by Horta.

So the exhibition was preceded by a collection that you built up and that is the collection which was shown in the courtyard, the collection of objects, and you invented, together with a guy whom I don't know called Czarnowsky, a kind of display feature.
Czarnowsky is a remarkable person. He was my classmate in Princeton and a very, very great organizer. He is the man who made it possible for the show to happen because he is an extraordinary producer, a very great producer. I designed the boxes and the idea was that the show would travel in the boxes with the flaps down and then when they were opened up they would be clean and the dirty face was to become hidden when flipped. When installed it looked like a supermarket at Carnac. I deliberately didn't give any hierarchy to any display; it was just a neutral grid, they were all placed on a neutral grid, and the inside of the museum was used for the newly produced environments.

Was that supermarket display association critical or was it parodistic in a way? Ironic?
Good irony is not intentional.

Very nice. That is very nice.
[Laughs]

The show in the courtyard with the display feature, the boxes and the collection, and inside you decided to commission those environments. So it wasn't about representation, it was actually about spaces which could really be experienced, a one-to-one thing.
The main thing is that the designers genuinely felt that designing an object did not exhaust in any way the possibilities of design. They believed they could prove much deeper, and that the Italian furniture industry should venture into making not only furniture but the whole interior of the house, bathrooms and kitchens and all the rest. You could see that idea implicit in the Zanuso environment, you can see it in Sottsass elements, you can see it also in certain Colombo elements and indeed they showed it. The only insurmountable fact in achieving such integration was that the unions wouldn't allow this to happen because, as per their rules, a building has to be built by a builders' union so there was no way that industry could upgrade itself from making tables and chairs to making a total environment which would then

be brought into an empty shell. But the idea was contained very strongly in these environments. I was able to bring it about because the museum provided the great institutional support; it was also a great enticement for the designers.

But I also exercised a role which I think the curator should exercise. I was the producer. I had one a case where a designer showed me something and I said, "Ha! Would you mind turning this Piscatorial stage around so I can have another 'Ha' because it's a very small little 'Ha' and it's a very stupid little joke. Go back to the drafting table and do something better." He demurred but he came back with something extraordinary. I had another designer who showed me something and I said, "Hey. You've got it all wrong. MoMA is not *Vogue* magazine; you can do something better, do something better." Then what I got from this person was, "No. You won't have me in the show then." I said, "I will have you in the show if I have to nail you to the drafting table and you will design." And she designed something splendid.

But it is basically a synthesis you made at the end; you brought it all together.
I had to bring it together. My conviction was that as a curator it was my task to bring up to a level of consciousness the ideas which are embodied in the objects or in the images that the designers were creating.

As for what you call Pesce's Prologue; he made an extraordinary commentary on society in general by proposing this environment that was an underground environment as found after an Earth-devastating war. I was madly lucky in getting the museum to let me remove the big service lift on which they carried up all the art and used the empty hole to create this environment so people would look at it from above. Pesce had a film made by a Swiss called Zugg, set with the "survivors" in the cave-like space, that was fascinating. Every other environment also had a film produced by the designer and I had written a number of prefaces and prologues to guide the people through this phenomenon. I was interested in showing that Italy was an extraordinarily lively culture, a society that believed that you make culture when you design a fountain pen. They also believe you can use designed objects to criticize culture. Nowadays, of course, the situation has evolved as follows: all those objects, environments, and ideas have traveled from the museum to the marketplace. They were the harbingers of things to come and that is what has now come about. Have they changed the society as we expected? Not really. But they gave us a great amount of pleasure; they kept us warm at night. At least they warmed my heart. And I think that any culture capable of such creations deserves a great badge of honor. We are now looking back thirty-five years after the show has taken place.

In the pro-design section there is a kind of optimism, one would say.
They thought they could make better products with a considerable content of sensuality so that the product could also appeal on an emotional level. Zanuso, Sapper, Roselli, and Bellini, among many others, were people who designed very beautiful objects to fit right into the existing culture. They did not get into contestation.

Joe Colombo was an absolute reformist. Reformists are people who understood their industry but always wanted to propose something that the culture was not yet ready to accept. But they felt that, "What do you mean? You can do it." At least one prototype was built. The capacity was there, the society wouldn't accept it, or the system of distribution or the allocation of resources.

Sottsass is a reformist?
Sottsass is a reformist and at the same time a contestatory.

He is a kind of anarchic reformist?
You can say that he is the elitist branch of the anarchistic movement, such an elitist branch is understandably composed of only one person, himself.

What did he do for your show?
He designed a splendid, an absolutely splendid, series of units on casters. One unit was a kitchen, another unit was a bath, the third one was a place to sit and the fourth one became a bed, and they were many — just imagine prisms of plastic connected one to the other and you could create a serpentine caravan of those prisms because they had wheels. If you lived in a loft, and that was, of course, Milan's idea of supreme modern life, you could use the elements and constantly reformulate your place every day. You could partition it as a palace one day or as a garrison the next. The idea was a very strong one, and the realization was very good. He, being what he is, deliberately painted them an ugly gray color. He told me he painted them this ugly gray so people wouldn't be too seduced by the object. What a melancholic ideal! He has a heavy constituent of pathos within himself.

And what about the mobile category with Zanuso and Bellini.
This is a category invented for ease of taxonomy (and we all know what comforts professors find in taxonomy). I had only three categories: the conformist, the reformist and the contestatory. Those designers fitted within the reformist slot.

We are not in a rush. Let's continue.
Let's continue or else you can say something. For example, people say, "Oh, you are doing this earth thing, this is untried architecture, this is a-tectonic." I want to remind them of, for example, of Buontalenti, Giambologna, and Francesco de Medici, 1560 to 1580, at Pratolino, outside Florence, where they built a garden with houses in the guise of hills and gods of the rivers. Of course the Renaissance had by then already exhausted its capacity, afterwards came the Mannerists, who felt themselves in a golden cage and could only demonstrate their revolt by using acid colors, as is the case with Sodoma, or yank columns out of their regular grid order as Giulio Romano did in the Palazzo Te. The people who afterwards wanted to break away from the Mannerists' golden cage had to become birds, they had to become natural entities, so they built this archinature in Pratolino. I think between 1569 and 1580. It was under the reign of Francesco de' Medici. Francesco de' Medici was later poisoned by his courtiers (that has been factually established) because he paid much more attention to artistic matters than to matters of state.

Imageries and also the illustrations here in the book, particularly the contestatory part, involved very spectacular installations. Seemingly Henri Lefebvre was the inspiration for you to define this contestation.
Yes.

And there are these incredible installations by studios and by architects which have become landmarks. Can you tell us anything about them?
I will do it at lunchtime. Let me tell you about the Spanish Catholic kings Fernando and Isabella. When they entered Cordoba, from where they had dislodged the Moors, they found writings on the wall in very beautiful Arabic calligraphy that were translated for them. The writings said, "Women are like shadows. If you follow them, they escape from you. If you ignore them, they run after you." So ignoring you has worked! [Laughter]

It's a great, fantastic, long interview. It is much better to do it on some exhibition. By the way, we have not talked about the exhibiton *The Taxi Project* yet, because after the *Domestic Landscape* exhibition you went in two directions: you did projects that had more to do with the production of reality, like *Taxi*, but also monographic shows like Barragán and the Barragán book you wrote.
Barragán and Pichler belong to a secret series of mine. Unknown to anybody this series was called "the unsung heroes of architecture." I thought that Pichler was a great architect, and I thought the same of Barragán. The other show I wanted to do very much was, of course, Amancio, but I had already resigned from the museum. I will now do Amancio as a book. I have all his drawings digitalized; I am like his ninth son. He had eight children and I am like the ninth in the family. Afterwards, I will do the exhibition; I will try to buy the drawings also so the family has some funds. Other shows I would like to do in the series are on Paul Nelson and Sergio Bernardes, a Brazilian architect and a man of utter charm and also a very good architect. A very interesting man. There were a number of those people.

So marginal heroes?
No. Unsung. The ones that didn't get any glory. The trumpets didn't sound for them, at least in this life.

So the unsung heroes was one series, then you went into the *Taxi* exhibition. The *Taxi* exhibition for me remains one of the most radical exhibitions in the history of MoMA.
You are too generous. Someone is now doing another show in New York on taxis, not at MoMA, and Paul Goldberger, who wrote for *New Yorker*, forgot completely about the one I did for MoMA and said, "MoMA should have done this type of show which is being done now." And of course, we had done it. The first hybrid car ever made was commissioned for that show. Volkswagen did it. It must be a German design, of course, they said, "Oh, we can't." I said, "Use Giugiaro to design the body." This was a hybrid car, electric and diesel, the first diesel engine Volkswagen had done and it was made by Ricardo in England, really. VW said, "Oh no! People will only look at the design by Giugiaro; we want them to look at the engineering" so they put this magnificent piece of anticipatory hybrid engineering in a VW camper. I said, "You are so bloody wrong. People will laugh when they look at the camper; they won't see the hybrid." Anyway, Giugiaro had produced a beautiful design for a taxi so I took it to Alfa Romeo and Alfa Romeo produced it, but it wasn't hybrid. So the first true hybrid ever was this camper and it ran for 45,000 kilometers in Volksburg with 4 normal batteries sitting under the back seat.

So the idea came a bit earlier in 1973 and we are back to the unsung heroes here as well because you start your Introduction by saying, "Taxis are the unsung heroes of urban transportation."
What people don't know is that in the United States, not in New York, of course, but in the small towns, taxis carry almost fifty percent of all the public passengers including old people who have to go to hospitals and do other things because there are no buses, so the taxis provide that type of service; they subsidize it for elderly people. So that's the reason I said that. But, you know, I am a man of few ideas many times repeated, so don't be so surprised you find me using the same words!

Still, the idea of the exhibition was very unusual because it was really about improving the world in a certain way by making taxis which would not be standardized but which would invent new forms of taxis. I was wondering how this idea came about and if you could tell me a bit more about that idea of the production of reality because the show is also a form of transition. One can say the *Taxi* show is clearly the moment when you then went into industrial design; it is a transition.
No, I was really keen; that was an idea I had and I was in a position to bring it about, to take it around. I thought the museum could provide the proper institutional support to bring that about. So I convinced the board of trustees and they said, "Yes, you can do the show, provided you have Detroit present in some decent way." Come on, you can very well figure out the board of trustees is a bunch of people who have Detroit and the good of America in mind. So I did the round of all the European manufacturers and I asked David Rockefeller, who was the chairman of the board at that time, to write letters to Gerstner, who was the president of General Motors, to the president of Chrysler (I can't remember the name) and also Ford, of course. Neither Ford nor Chrysler were interested to even answer him (not me). Gerstner of General Motors answered him, "Hey David, why don't you take care of your bank, I don't think it's going so well. We sell fifty thousand Chevrolets as taxis and no one has complained so we don't have to do anything special; they serve the purpose of the city, blah, blah." When I wrote the catalog I said GM told in few words the Museum to get lost.

So naturally the next thing we hear, GM's lawyers are talking about suing MoMA for libel. So I called up their lawyer and I said, "You don't want to sue us for libel because here is the letter your president sent." He didn't want to participate; he said that we shouldn't bother with taxis and that is exactly what the catalog quoted. We made very sure we quoted him completely so they couldn't claim that statement was taken out of context or anything of that sort. I only withheld his comments about the Chase Manhattan Bank but kept his comment about our cities and its drivers not needing another type of taxi service since a Chevrolet is good enough for everybody. So we had the taxis made by VW, Volvo, and Alfa Romeo, but I couldn't get Detroit to participate. Vice President Ford who then became President Ford, knew me because I had designed the Museum of Grand Rapids when he was the representative for Michigan State. When he became President, after Nixon's resignation, he met me once on some occasion and asked me, "How is the taxi project going?" I said, "We have problems because Detroit doesn't want to participate." Detroit, being in Michigan, was one of his constituencies. He said, "What will be needed?" I said, "Probably they need a subsidy." A few days later the Department of Transportation called the museum and they offered two million dollars for a subsidy. Of course it was money from the state, and it had to be made available publicly, so we published a series of articles in newspapers and wrote to every manufacturer of cars saying this amount of money is available; if you want to participate there are two million dollars. Chrysler answered so I was very happy. I took the plane to Detroit, got a taxi driver who was completely drugged, so I had to ask him to sit back and I drove. Somehow I found Chrysler—it was quite something. They said, "We will make it for you." I said, "Fine. What chassis will you use?" They said, "No. We will make it in gypsum, *scagliola,* the traditional one-to-one scale model." I said, "I think it will be marvelous. When people open up the door of the Alfa Romeo they will have a chance to look at the interiors. When they try to open your door they will have the door handle in their hands." That was the end of our meeting. They wouldn't take the money. They didn't want to do anything. So I went back to the board of trustees and said, "This is the situation. Volvo is doing it, Alfa Romeo is doing it, Volkswagen is doing it, the two million are there, two small little companies, one of them burning grain for alcohol is willing to do it (it was pretty ugly but they were willing to do it). What shall I do?" They said to go ahead without Detroit.

Those vehicles were designed to be tested by the Department of Transportation, so the vehicles were given to the Department of Transportation; each of the manufacturers had produced two working prototypes, one to test and crush to prove they were roadworthy, and one for the show. But there was something very amusing. I wanted the union of taxi drivers to participate in the show and they wouldn't do it, so I got someone who had great relations with the unions, to get me an interview with Harry van Arsdale, who was president of the union of taxi drivers. He told me the real reason for his union's reluctance to participate was that the union of taxi drivers was a great investor into taxi fleets so they were the ones who didn't want to change taxis; they didn't want to change anything. They couldn't care less about the needs of the drivers. They had a great investment in the fleets. They were defending the owners not the drivers. I told him, "Mr. van Arsdale, do you know something? In Argentina, where I was born, every union has its hospital so it was noticed by the union of taxi drivers' hospital that they had an above average number of complaints about sexual impotence. When they looked into that they came to the conclusion that that problem was due to the fact that the men were sitting there all day long in the taxis, they didn't get up, there was a retention of fluid and therefore they couldn't perform." He said, "You're a son of a bitch." I suggested to him, "It takes one to know one. You understood me? You want that story known? You don't.

Well, I want just your sponsorship, just to say that the union is a sponsor of that. Not money, nothing else. I can't do it without the union being part." So his revenge was that before we opened to the public, the television came to film the show. He had sent two goons (goons are like gangsters) who got somehow into the Museum screaming in front of the TV cameras, "This prototype, if it went into a tunnel it would collapse and if it went into a washing machine it would fly away and if it went into a bridge the wind will take it," all these types of things. Of course the television people understood immediately what was going on and did not record it. They really looked as if they had been made in Hollywood to look like gangsters. But that was the situation that we had.

After the show closed the taxis were tested by the Department of Transportation in Washington. I had at that time convinced the Department of Transportation to do something I pretend was intelligent, that is to lend money to people who would buy the taxis which were produced along the lines of the taxis the Musuem presented. Here was Detroit having a crisis, there were a tremendous number of suppliers for the car industry with great problems. People who made the roofs of convertible cars had no work. Why not have them make a taxi which is specifically made to be a taxi and the Department of Transportation make loans to facilitate their purchase by independent taxi drivers? Imagine that you make available a budget of 300 million dollars for loans and give a loan of $30,000. That means you have ten thousand taxis out. We put outside this taxi's door the Department of Transportation's emblems, so that it gets deserved kudos. This idea would have certainly helped the Detroit industry; the taxi workers could afford to buy these types of taxis, and why not bring some recognition to DoT. I am a very practical sort; it is of no use to have a show if I can't bring about to the streets the actual taxis. So that was the idea and everything was to go in that direction until everything changed because a new Department of Transportation head was appointed and it all fell by the wayside.

Were there any unrealized projects at MoMA, any exhibitions they wouldn't let you do, which were censored?
No, they wouldn't have been censored. I don't remember, frankly. If there ever was one—and I would have to sit and think, maybe I have repressed them—I can't at this moment remember one.

Have there been unrealized projects in your life? Every architect has unrealized projects but I was wondering if you have unrealized projects that are particularly dear to you that you would like to see realized.
Universitas. That's the one project, the big regret in my life, the *Universitas.* I would have wanted very much to reformulate the notion of the *Universitas.* My ambitions are very small, as you can see! That's the one that weighs on my head and I am at least happy now that MoMA will publish the proceedings; we have papers and comments by Lefebvre, Hannah Arendt, Jean Baudrillard, and thirty more luminaries. We had Manuel Castells's contribution. I don't know if you know him. Manuel Castells still had his mother's milk under his nose, he was still so young.

When did you met him?
I read something by him. I also discovered Susan Sontag, not that she needed me to discover her.

I know that you also discovered Shigeru Ban.
Yes, also Barry Bergdoll and Michael Sorkin, as I told you before. Shigeru was a student at Cooper and he came to see me and I didn't know really what he wanted. He was in the last year at Cooper, I think. I developed great trust in him and I liked the way he was thinking. When he went back to Japan he was appointed curator of Axis Gallery in Roppongi, that was something sponsored by the Ishibashi family of the Bridgestone Company, who make tires. So one of the first shows he did was a show of my work in 1985 and since then he has been becoming a better and better architect. I place great stock in such discoveries. For me to have discovered them gives me great joy.

That is fantastically covered. Thank you so much. There is one last thing which is a mystery to me. In your Museum of Modern Art curatorial activity there is also some design because you obviously designed the catalog or the poster of the *Taxi* project and the *New Domestic Landscape,* but you also designed some strange cards. There is this "Temporarily removed by order of the curator" card, which is really, really funny and I wanted to ask you where this was produced.
It was sold by the Museum. And there was another card — I don't know if it's here — which I like even more than that.

Designed by you?
Yes; that's it, that one in the book. It was a piece of tracing paper, vellum, with a sunflower printed on it; you fold it in nine pieces in a certain way and there is a little cutout here where you affix the stamp and the stamp works like a seal. That card contained sunflower seeds and people could either eat them or plant them.

Fantastic!
That was 1974; it is more than SIXTY years old. I also did calendars for MoMA. This one I am showing you is an interoffice envelope, the prototype version.

And that is the deluxe version.
See it here. Every month is a large interoffice envelope. So people can use the monthly envelopes to keep their memories inside; things like tickets, a hair lock, anything they want, and of course they can pull away the die-cut days and as they pull away the days the envelope becomes transparent in a certain way and you can see what is inside so you have all your memories visible.

Great. Thank you so much.

Italy: The New Domestic Landscape

Exhibition, 1972

The exhibition *Italy: The New Domestic Landscape,* held in 1972 at the Museum of Modern Art and curated by Emilio Ambasz, brought to light systematic conflicts embodied in the current state of design, art, and industry. By using the state of Italian design as an example — in the best sense of the word — the exhibition brought together industries and schools as diverse as the automotive and furniture industries, engineering and historical fields, art nouveau experts and futurists. The project asked the question: How does the industrialization of design and production make a new domestic landscape — with regards to particular objects and people, the overall layouts of homes, cities, and landscapes, and society as a whole? Some of the overarching themes are embodied in the poster and the book cover of the accompanying publication, designed by Ambasz. They allow an ever-changing array of objects to move in place, relate to one another, and build new landscapes. The exhibition ran from May 26 to September 11 and was one of the most ambitious ever undertaken by MoMA.

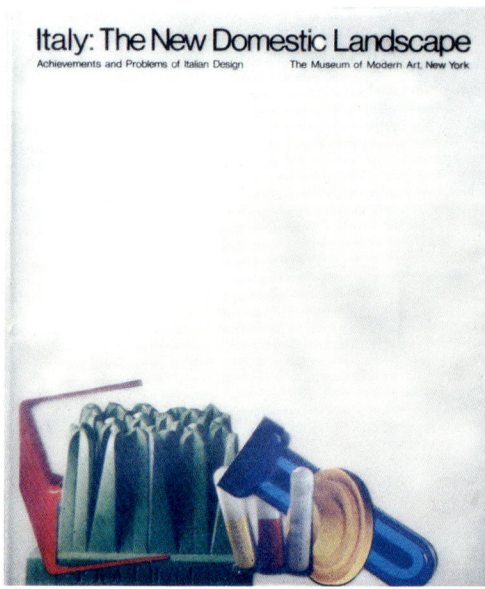

Exhibition catalog, 1972

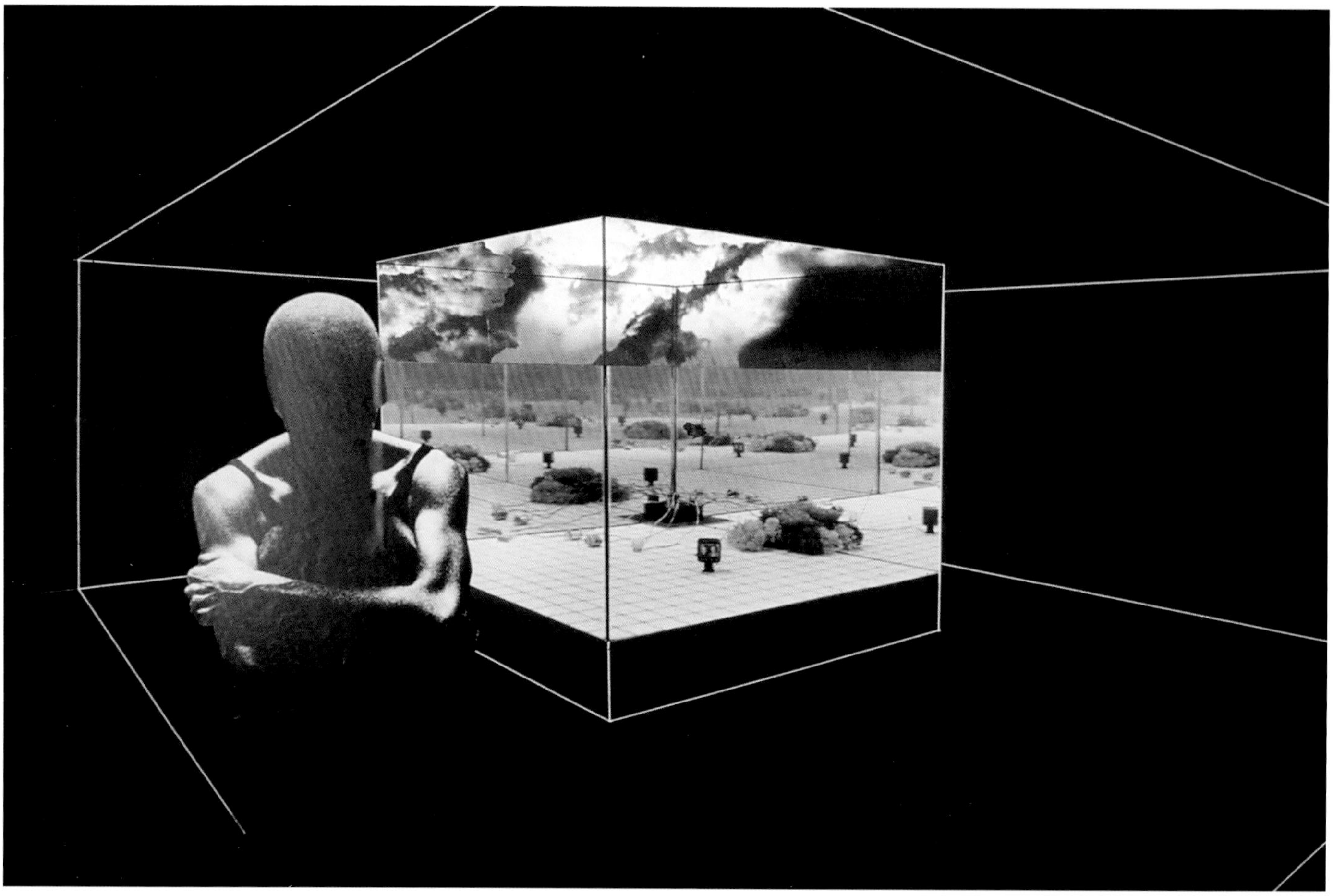

Superstudio, *Supersurface*, 1971

Ettore Sottsass Jr, environment section from the exhibition *Italy: The New Domestic Landscape*, 1972

Gaetano Pesce, *Project for an Underground City in the Age of Great Contaminations*, 1972

Mario Bellini, Kar-a-sutra concept car, installation view of the exhibition *Italy: The New Domestic Landscape,* 1972

The Taxi Project: Realistic Solutions for Today

Exhibition, 1976

During his time as the curator of design at the Museum of Modern Art, and soon after the onset of what came to be known as the energy crisis of the 1970s, Emilio Ambasz started an innovative project called *The Taxi Project: Realistic Solutions for Today.* Combining an exhibition, prototypes, and a publication, Ambasz sought to bring together car manufacturers, public transportation agencies, urban theorists, historians, and designers to produce a better taxi. Renowned manufacturers such as Volvo, Volkswagen, and Alfa Romeo submitted working and tested prototypes for the exhibition—all put in comparison to the traditional London taxi. Designs ranged from the same-old to the all-new, and the motors encompassed types as varying as hybrid, diesel, and steam engines. The prototypes—as required by the contest—were evidence of increasing efforts to reduce pollution, accommodate the handicapped, and improve the experiences of both drivers and passengers. The exhibition ran from June 16 to September 7 and achieved major success.

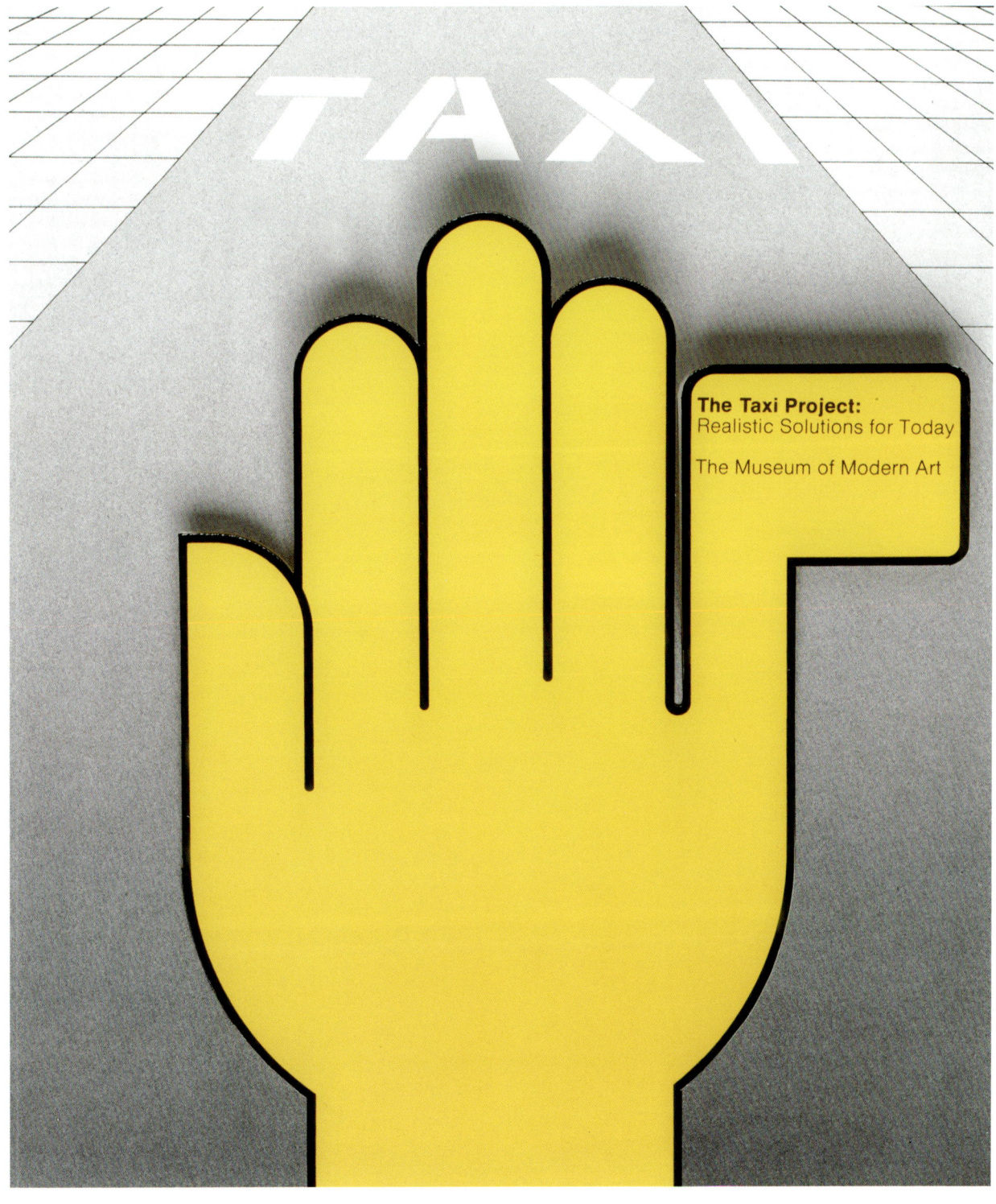

Ambasz's architecture is at its best when it is largely a landscape or a waterscape, when it is subsumed by the topographic — a situation in which earth, water, and vegetation take precedence over architecture, with the result that architecture begins to disappear.

Kenneth Frampton

THE ARCHITECT AS IDEAS MAN: A MEMOIR/CRITIQUE

Kenneth Frampton

I met Emilio Ambasz on my first visit to the US in the fall of 1964. He was the first person I encountered, on the occasion of an architectural conference in Princeton University, in the service of which he had been assigned the dubious honor of welcoming me at Newark Airport.[1] The following year saw me serving as visiting tutor in the school of architecture at Princeton, where I had the privilege of having Emilio as one of my students, although I can hardly claim to have contributed much to his capacity as a designer. Having been a teenage protégé of the distinguished Argentine architect Amancio Williams, his stature as an architect was already quite developed, as was evident from his thesis project, a design for the National Library in Buenos Aires. Like many of Williams's tectonically audacious proposals, Ambasz's national library was projected as a cantilevered *boite à miracles* supported well clear of the ground on four giant pylons.

Ambasz passed from graduating in the spring of 1966 to become a member of the faculty in the fall of 1967, with the new dean Robert Geddes rationalizing this appointment on the grounds that Ambasz was a precocious genius who had totally bypassed the slow learning curve of young adulthood to emerge as a junior master overnight. Although Ambasz was on the faculty at Princeton for barely two years, he was there long enough to have a crucial influence on the discourse of the school. It was certainly Ambasz who saw to it that Tomás Maldonado, the former director of the Hochschule für Gestaltung, Ulm, summarily shut down in 1968, was invited to Princeton to serve as visiting professor in the school.[2] Ambasz was also instrumental in having the French sociologist Abraham Moles of the University of Strasbourg, invited to the school to give a few seminars on the sociology of objects.[3]

Ambasz's next meteoric move in 1969 saw him become curator of design at the Museum of Modern Art, New York, a position which Ambasz would occupy for seven years before deciding to branch out on his own as architect/designer in private practice. Before giving a brief account of Ambasz's tenure at MoMA. I would like to convey something of his modest yet stylish modus vivendi during his initial years in Manhattan—his small studio apartment in an Art Deco building on Central Park West, which he has continued to maintain as a pied-à-terre in the interim. Of this white interior, empty save for one or two elegant pieces of furniture, three key elements remain engraved in my memory—a Castiglioni Boalum lamp snaking along the floor, a talismanic woolen wall hanging by the Chilean textile artist Sheila Hicks, and a distant, spectacular view of Frank Lloyd Wright's Guggenheim Museum, seen on axis in half light, across Olmsted's Central Park, as a brightly illuminated paradise pavilion reflected in the surface of the reservoir.

As Ambasz frequently insists, the mythic has always played a major role in his design work, irrespective of whether the object under consideration is a building or an industrial product. Something of this is already evident in his early forays into the closely linked fields of graphics and packaging, such as the highly original black-on-black, three-dimensional "peeling" poster that he designed for an exhibition of Geigy products staged in the School of Architecture, Princeton University around the time of his graduation. A similar origami play is present in his early prototype for a disposable cardboard razor which when packed flat assumed the form of a standard book of matches but which, when opened up and folded along prescored lines, transformed itself into a cardboard handle of triangular section and a preformed stiff cardboard head into which a thin metal razor strip was embedded.

Ambasz first revealed himself unequivocally as a generic intellectual in the four lectures that he gave in the Hochschule für Gestaltung in 1968, the first of which was published in *Perspecta* 12 in the following year. Although this lecture, entitled "Notes Toward the Formulation of a Design Discourse," may be seen in retrospect as echoing the objective design methods of the moment, particularly as elaborated by such figures as the peripatetic German mathematician Horst Rittel, Ambasz was simultaneously critical of both traditional studio teaching and the supposedly objective systematic new design methods. With regard to the latter he wrote:

> Such approaches have so far, with some distinguished, though rare, exceptions, given rise only to unsophisticated rigid "method-idolatry," or to an opportunistic embracement of design methodology as a design panacea.[4]

It is typical of the period that the traditional terms *architecture* and *urbanism* do not appear in this text and that in their stead the author employs the more comprehensive and seemingly objective term *environmental design*. At the end of this *Perspecta* essay, Ambasz's ingenious play with the terms *archetype, prototype, type,* and *stereotype* representing the processal rise and fall of a received type-form over time, seems to anticipate the trajectory of his own career as his work has oscillated back and forth between the archetype of myth and the prototype of industry.

In the short passage of time between 1970 and 1972, Ambasz brought the Museum of Modern Art, in its role as the ultimate arbiter of cultural modernity, to confront the shifting sands of modernism on two fronts at once; in the first instance through the *Universitas project* staged at MoMA on January 8 and 9, 1972, which engaged some twenty intellectuals in debating the possibility of establishing a university of design, and in the second, the exhibition entitled *Italy: The New Domestic Landscape,* which was displayed in the museum and its garden from May 26 to September 11 of the same year.

It is difficult to convey the unique and exceptionally ambitious character of Ambasz's *Universitas Project,* which not only envisaged the creation of a third type of university comparable to the successive humanist and technoscientific universities of the past, but also one which was implicitly predicated on a third way of nondeterministic, technoecological development transcending what was then still the historical standoff between the communist and capitalist worlds. The subtle and complex epistemological scope of his highly original postulation of a university of design may be most succinctly characterized by citing a salient passage from the MoMA Working Paper compiled in 1971, with the express purpose of establishing an agenda for the *Universitas* debate. The following is from the so-called Black Book given to all the participants before the event:

> The determinism of natural science, not allowing man the ability to alter by the choice of his actions, the course of things, excludes a concern with ethics. Design which is based precisely on man's possessing that ability must by the same token, give ethics a central place in its thought and cannot profitably separate facts from value as science does. […] The future, it has been cogently argued, can differ radically from the present only if it is based on a different set of *values,* to truly create the future and not just extend the present, a change in the overall configuration of values must be brought about. In view of those who attempt to develop a normative approach to design—in contrast to those who conceive it in empirical terms—it follows that the examination of values should be the main concern of design.[5]

Owing to space limitations it would not be appropriate to elaborate here on the pros and cons of the ensuing *Universitas* debate. Suffice it to say that there seems to have been something of a consensus that one of the most cogent difficulties to be overcome in the formulation of any new value-laden design discourse was the delicate balance to be maintained between academic freedom and the instrumentality of power.

In terms of the exhibited material, *Italy: The New Domestic Landscape* did not overtly initiate any such extended debate over values *per se* except insofar as Ambasz arranged the exhibits according to the following discrete categories. Firstly, objects selected for their formal and technical attributes; secondly, those selected for their sociocultural implications and, thirdly, those grouped together by virtue of their providing "for more flexible patterns of use and arrangement."

Where the first section was in essence an anthology of the best of Italian industrial design of the previous decade (1961–1971), the second section celebrated the emergence of an ironic, nonfunctional postmodern sensibility which made itself manifest through rather grotesque enlargements of everyday objects, such as the famous baseball mitt sofa of 1970 or Ettore Sottsas's Pop Art-cum-Neo-Art Deco furniture pieces that bordered on ludic kitsch. At the same time, the third section, devoted to recombinatory permutable forms was almost indistinguishable from the larger environmental set pieces, commissioned specifically for the show.

The latent focus of this exhibition was the Italian *architettura radicale* movement which seems to have been inspired by Mario Tronti's Neo-Gramscian concept of a resistant stance that was at once "both inside and outside capitalism." This sensibility first rose to the fore in 1966 with the formation of the Superstudio group in Florence, along with the equally radical Gruppo Strum, who staged an exhibition entitled *Utopia &/or Revolution* in Turin in 1969. Where the monumental panoramas of Superstudio depicted a pacified world without objects ("There will be no further need for cities or castles; There will be no further reason for roads and squares") Gruppo Strum produced newssheets featuring semifictive fottoromanzo accounts of direct political action appropriating both the means of production and existing residential stock. Thus, as Felicity Scott has put it in her analysis of the exhibition; "(a) powerful allegorical procedure can be traced throughout the exhibition, from the image/text relations to the audiovisual sessions to the installation strategies, recasting curatorial and institutional practices with a political force not to be repeated at MoMA."[6]

It is significant that Ambasz's last two exhibitions at MoMA, both staged in 1976, would spell out in categorical terms the two sides of his creative and professional persona; on the one hand his ultrasensitive awareness of the poetic potential latent in both architecture and landscape as this will be especially evident in his pathbreaking exhibition on the work of the Mexican architect Luis Barragán, as represented through the photographs of Salas Portugal; on the other, the *Taxi* project, in which he exhibited four taxi prototypes especially designed to fulfill the task of serving the greater New York metropolitan area.

Of the two it is this last that returns us to that side of his creative personality that is most connected to his innate capacity for invention. In this regard, his role as Chief Design Consultant for Cummins Engine Corp. a position to which he was appointed in 1978 by Irwin Miller, is exemplary. Early on in this position Ambasz came up with the idea of cooling the standard diesel engine with oil, with the proviso that the oil, in its turn, would be cooled outside the block with the water running inside a hermetically sealed container. It was this breakthrough that gave him the authority to influence the overall block profile and shape of the N14 Cummins diesel engine and its attendant variations.[7] Ambasz would hold this advisory position for some twenty-five years, while devoting the remainder of his time to a double-headed transatlantic practice of architect/industrial designer divided between Bologna and Manhattan.

If the N14 engine was a breakthrough as far as his consultancy was concerned, an event of equally decisive importance with regard to his subsequent career as a designer was the Vertebra Chair of 1974, the world's first articulated chair designed in collaboration with the Italian designer Giancarlo Piretti. This was an office chair designed to respond to the sitter's every movement, without having to adjust the angle and profile of the seat through mechanical means. Ambasz would follow this chair with two subsequent chairs designed with Piretti—the almost equally responsive Dorsal Chair of 1978 and the Lumb-R Chair of 1981. The Vertair Chair of 1991, designed by Ambasz alone, recasts the Vertebra Chair, to arrive at a more continuously flexible, luxurious form which, by virtue of being upholstered with leather-elastic banding gave to the chair a "soft-tech" look that in many respects would become the ethos of Ambasz's style as a product designer.

In his pursuit of a soft-tech ergonomics Ambasz took as his parti pris the Italian anti-Bauhaus design tradition as this was possibly first fully exemplified in Gino Sarfatti's table lamp of 1965, in which a cylindrical lamp shade was mounted on top of a diminutive artificial leather bag filled with pellets so that the angle of the lamp could be adjusted by manipulating the pouch. This radical concept attains its apotheosis in the world famous "bean bag" chair of 1969. Ambasz's own excursus into a similar soft-tech aesthetic came with his brightly colored and ludic Flexibol pen of 1985, comprised of two barrels connected by a semistiff ringed tube that would bend in the pocket to accommodate the movement of the body. Once the lower barrel was twisted, however, the two halves would snap together to form a rigid shaft and reveal the nib. As Eric Chan, who worked with Ambasz on this ingenious design has remarked: "It's a very poetic idea with a distinct before and after. Emilio always works to simplify a mechanism not in a mechanical way but in an organic way. He's very good at nontraditional ways of solving problems. [...] For Emilio an idea has to be groundbreaking or it is not worth doing."[8]

Soft tech became an intriguing preoccupation in Ambasz's so-called Soft Series of 1990, successively comprising a portable radio-cassette player, a soft computer notebook, and a soft "handkerchief" TV. In each instance a rigid shell, incorporating the electronic component, was embedded into a padded, flexible leather case, which in the computer notebook and the TV could be unfolded in such a way as to establish the angle of the screen in a convenient position. As with the Veterbra Chair the soft vs. hard paradigm was integrated into Ambasz's activity as a furniture designer across a wide range. This is evident in Qualis Office Seating (1989) where the electrically welded, metal-framed pieces are upholstered with a padded cover that may be readily unzipped from the frame for cleaning or replacement.

Throughout Ambasz's prolific output as a product designer, a balance is maintained between the *inventive,* the *ludic,* and the *normative* and it is this tripartite equilibrium that distinguishes his comprehensive vision as an architect from his precise ingenuity as a designer. It could be said that where his product design is necessarily ergonomical, his architecture is ostensibly ecological in as much as the dialogue between hard and soft in his architecture is divided between the inorganic rigidity of built form and the organic flexibility of plant material—the prevalence of "green over gray" as he would put it. As in all architecture, the common ground sustaining the two is the earth itself, even to the extent of allowing the vegetation to engulf the building in its entirety. At the same time, the building, as opposed to the *produktform*—to coin Max Bill's felicitous phrase—is never a free floating object but is instead firmly anchored into the ground. One may also note in this regard that the *normative* tends to be absorbed by the *mythic* in Ambasz's architecture. It is this perhaps that accounts for Ambasz's attitude toward the realization of built form, as opposed to the *produktform* where the designer is so implicated in the processes of production as to be virtually inseparable from the toolmaker; that is to say the "designer as producer," to paraphrase the title of Walter Benjamin's essay "The Author as Producer" of 1934.[9]

To a large extent we may say that Ambasz as architect is Ambasz the mythmaker and herein resides a certain ambivalence, since nothing is more alien to myth than realization. This schism between the ideal and the real recalls Louis Kahn's aphoristic lament put into the mouth of an apochryphal student to the effect that:

> I dream of spaces full of wonder. Spaces that rise and envelop flowingly without beginning, without end, of a jointless material white and gold. When I place the first line on paper to capture the dream, the dream becomes less.[10]

Ambasz's architecture is at its best when it is largely a landscape or a waterscape, when it is subsumed by the topographic—a situation in which earth, water, and vegetation take precedence over architecture, with the result that architecture begins to disappear. Thus, as he has revealed on more than one occasion, his primary aim has been to eliminate architecture, to produce that which Fumihiko Maki has perceptively recognized as a *meta architecture*.[11]

It is just this meta quality that makes some of his projects so compelling. Thus, within his prodigious output as an architect, particular works rise to the fore as ineffable high points that seem to lie, at times, uncannily close to the Miesian ethos of almost nothing *beinahe nichts*. Two examples are his Cooperative of Mexican-American Grape Growers projected for Borrego Springs, California, in 1976 and his Center for Applied Computer Research, designed for Mexico City in 1975. Where the first is partially rendered as being subterranean, as in the case of the excavated chapel or, alternatively, is conceived as a provisional encampment and virtually invisible by being set beneath a seemingly infinite trellis of vines, elevated well above grade and stretching out to the horizon, the second is represented as floating upon an *alto plano* lake in the Aztec tradition. Ambasz's description of this last posits a pre-PC narrative in which there is a significant interplay between topography, technology, and the potential for a liberative, protosustainable technique.

> The site conditions suggested the solution to the need for flexibility. Because Mexico City and its surroundings are built on the landfilled site of an ancient lagoon, the building has a large (180 meters square) water basin which drains the soil and prevents foundation problems. To take full advantage of this basin, the building's office/workspaces have been designed as barges, which float until positioned in place. Their watertight compartments are then filled, and the barges come to rest on the bottom of the four-feet-deep basin. To reposition them, water is pumped out of the compartments and the barges are floated to new locations.
>
> The premise behind the design of this flexible environment is that nobody should have to work. At worst, it should be possible to work at home, in which case the need would not be for a large building but for a relatively small one to house the computer and receive messages. The building has been conceived, therefore, as a set of elements that could be progressively removed as the user's preferences change and the need for on-site space diminishes. Ultimately, only the silent walls and a single barge, turned into an island of flowers, would remain.[12]

Strangely enough, this description of 1988 does not mention that which the model patently shows (along with a Boullée-like vaporous cloud rising off the lake), namely, that the energy for the center would have been provided by a forty-meter-long solar collector wall, inclined toward the south. Here one is witness to the transformation of a tectonic invention into a mythic narrative. Something equally tectonic, if less mythic, obtains with Ambasz's 1979 proposal for a thirty-story tower to be built on top of the American Folk Art Museum in New York. Although this is an entirely "hard" structure, quite atypical for Ambasz, the logic of the invention is nonetheless meaningful for the way in which it synthesizes three typologies at once; firstly, the bottom, middle, and top of the early skyscraper; secondly, an inversion of the set-back, high-rise profile mandated in the 1920s for Art Deco skyscrapers, and, thirdly, an archetypal portico capable of providing a monumental entry into a prominent cultural institution. There would have been other incidental attributes had the work been realized, as the following description makes clear:

> The blocks (of offices) are set forward as they ascend, expanding their horizontal span as their supporting piers grow taller and thinner. The arrangement of the blocks reduces the tower's perceived mass; increases its resistance to wind pressure, allows greater penetration of light through the building, and expands the function of its interior while still presenting a unified face to the street.[13]

American Folk Art Museum, New York, USA, 1979

After this exercise in Manhattanism, Ambasz's architecture will assume a predominantly "green" cast in as much as his finest work of the 1980s and thereafter will be informed by an all-pervasive landscape character as found in the green tessellated public plazas projected in 1982 for Houston, Texas, and Salamanca, Spain, and in the semisubterranean Schlumberger Research Laboratories projected for open land near Austin, Texas, as well as in the partially buried botanical garden, the Lucile Halsell Conservatory built outside San Antonio, both of which date from the same year.

Ambasz's 1986 winning competition entry for the design of a master plan for the Universal Exposition, which was eventually staged in Seville in 1992 without following Emilio's design, was initially thought out not only as a waterscape but also as a comprehensively economic and ecological game plan capable of responding to the challenge and cultural significance of creating a large-scale commemorative exhibition. The project description says everything in this regard: "The master plan proposes three large lagoons on which most of the activity takes place. All of the exhibition pavilions are floating and after the exhibition finishes, they can be taken away leaving only a magnificent garden park. [...] The water symbolizes the indispensable communications link between Spain and the New World."[14]

Master plan for the Universal Exposition, Seville, Spain, 1992

A similar paysagiste stratagem is present in the best of Ambasz's architecture to date, including the handling of large megastructural complexes in such a way as to simulate the appearance of massive natural outcrops covered in vegetAation.

Throughout Ambasz's career there is a sense in which his work is permeated by a fundamentally schismatic identity. This much is suggested by his preoccupation with the idea of the double, in which the one cannot survive without the other and vice versa. This notion has never been more ironically expressed than in his exceptionally revealing essay of 1988 appropriately entitled, "I Ask Myself" wherein he wrote "As I often tell my friends, 'Emilio Ambasz' has two seemingly paradoxical selves within him: 'Emilio' is a solitary, cheerful man, anguished nevertheless because he hopes through his architecture to be welcomed by angels; 'Ambasz' is a sociably sad man anxious because he wants his products to be well received by men. I, on the other hand, am cheerfully sad and socially solitary."[15]

The poignancy of this introspective awareness surely goes some way toward accounting for the split between Ambasz's position as an inventor/industrial designer of readily achieved world stature and his identity as an architect caught between projects of tectonic lucidity and resolution and the much more rhetorical megapropositions that remain uncertainly suspended between the built and the unbuilt, between nature and culture, invariably yielding panoramic images of a discernable metaphysical character.

In addition to this schism is a cosmopolitan identity suspended between his precocious early career on the East Coast and his subsequent move to Italy and the ensuing seasonal commute between Manhattan and Bologna, coupled with the habitual fluency with which he passes from his native Spanish to English and Italian. Over and above such overlapping polarities as Emilio vs. Ambasz or the Northeastern seaboard vs. the Mediterranean, there remains the unifying persona of the peripatetic intellectual who is equally torn between the discourse of theory and the poetry of allegory. One may think of this last as the legacy of Walter Benjamin, to which Ambasz seems to have been susceptible some years before the first translations of Benjamin's writings into English. Relevant to this is the illustrated version of his "Anthology for a Spatial Buenos Aires" of 1966, a hand-bound copy of which I had the honor to receive from Emilio in Princeton, during the year in which it was compiled. Seemingly aspiring to Benjamin's ideal of writing an essay composed of nothing but quotations, this appraisal of Buenos Aires as an infinitely gridded utopia running out into the infinite expanse of the pampa was much indebted to Jorge Luis Borges, who opened Ambasz's collection of citations with the words "It seems to me a tale that Buenos Aires ever started. I judge her eternal as the water and the sky."

Elsewhere in this anthology we read another quote of Borges: "The hour of Buenos Aires is the afternoon, the hour of the desert. It is then that the city acquires its cosmical aura." Later still, "I want to talk about the plazas. In Buenos Aires the plazas — noble pools overstocked with freshness, congresses of patrician trees, stages for romantic rendezvous — are still waters where the streets resign their persistent geometrical flow and joyously disperse."[16] These evocative words of Borges placed side by side with equally lyrical passages from Le Corbusier's *Précisions sur un état présent de l'architecture et de l'urbanisme* of 1930, is rounded out at the end, by Martínez Estrada's prose poem *Civitas,* in which he elects to see the Argentine capital as a realized ideal city.

No subsequent text by Emilio would come close to portraying the promise of the modern project in such evocative, positive terms, not even his "Manhattan Capital of the 20th Century" of 1969, which he regarded as such a seminal résumé of his philosophy as to include it in the introductory material of both the *Universitas Project* and *Italy: The New Domestic Landscape.*[17]

Of all the numerous texts that Ambasz has written — his so-called *Working Fables for Skeptical Children* — one stands out as the most cogent statement of his complex political position, his essay entitled "Coda: A Pre-Design Condition," written in the same year as the student revolt of May 1968 by which it was inspired. At the end of this he writes:

In the past, when conflicts were more clearly defined along class lines, courses of action that could be taken to bring about change were

1 This was the occasion of the inaugural meeting of the Committee of Architects for the Study of Environment (CASE) which took place in the Lowrie House, Princeton University, in March 1964.
2 Tomás Maldonado, Argentine painter and design theorist. Maldonado was a strong influence on Ambasz's initial development. In his essay in *Perspecta* 12, Ambasz cites a particular seminar that Maldonado gave at Princeton in Fall 1966, entitled "Man & Environment."
3 Abraham Moles's most influential work in English was *Information Theory and Esthetic Perception* (Urbana: University of Illinois, 1963).
4 Emilio Ambasz, "Notes Toward the Formulation of Design Discourse," *Perspecta* 12 (1969): 58.
5 *The Universitas Project: Solutions for a Post-Technological Society,* was conceived and directed by Emilio Ambasz (New York: Museum of Modern Art, 2006), 21, 31, 32.
6 Felicity Scott, "Italian Design and the New Political Landscape" in *Analyzing Ambasz,* ed. Michael Sorkin (New York: Monacelli Press, 2004), 118.
7 Ambasz's own account of his relationship to Cummins Diesel is given in a letter to the author dated 29 June 2011. Ambasz succeeded the late Elliot Noyes in this position.
8 See Peter Hall, "The Immortal," in the present volume, 207–18. The Chan quote is from an interview with Peter Hall, September 20, 2002.
9 See Walter Benjamin, "The Author as Producer (1934)," in *Walter Benjamin: Selected Writings,* ed. Michael W. Jennings (Cambridge, MA: Harvard University Press, 1999), 768–82.
As Peter Hall puts it in his essay "The Immortal," 209 (see note 8): "Ambasz no longer expects commissions to come to him, instead he develops his products himself. With his Vetebra Chair of 1976, he sold a part of the future business to toolmakers, die-casters and upholsterers, persuading them to make tools and dies at no cost in anticipation of a share of the profits."
10 Louis Kahn, "Form & Design (1960)" in *Louis Kahn: Essential Texts,* ed. Robert Twombly (New York: Norton 2003), 62–63.
11 See Fumihiko Maki, "Primary Architecture," in *Emilio Ambasz Inventions: The Reality of the Ideal* (New York: Rizzoli, 1992), 74.
12 See Emilio Ambasz, *Architecture and Nature/Design and Artifice,* 2nd ed. (Milan: Mandadori/Electa, 2010), xliv and xlv.
13 Ibid., xliv.
14 Ibid., lviii.
15 Emilio Ambasz, "I Ask Myself," in *Emilio Ambasz: The Poetics of the Pragmatic* (New York: Rizzoli, 1988), 25.
16 Emilio Ambasz, "Anthology for a Spatial Buenos Aires (1966)," in *Emilio Ambasz: The Poetics of the Pragmatic,* 33–37 (see note 15).
17 Emilio Ambasz, "Manhattan Capital of the XXth Century (1969)," in *Emilio*

Emilio Ambasz and Kenneth Frampton, ca. 1968

Ambasz: The Poetics of the Pragmatic, 38–39 (see note 15).
18 Emilio Ambasz, "Coda: A Pre-Design Condition (1968)," in *Emilio Ambasz: The Poetics of the Pragmatic,* 49–51 (see note 15).

more narrowly determined. Whereas the pre-design structure that we contemplate would make action possible in a less determined situation, offering a broader choice of alternatives that exist as real possibilities and giving individuals more power to influence historical development. This social ethics should be based on the needs and aspirations of individuals. The child of an ethical decision, this pre-design structure should itself be a matrix of ethical decisions.[18]

Everything is summed up here in a few words, and recent writings would suggest that Ambasz's political credo has remained essentially the same over the forty-three years that have elapsed since then. The same fundamental ideas are seemingly adhered to — an emphasis on the ethical rather than the political as a strategic principle, a conviction that class struggle, in the twentieth-century sense of the term, with its former potential for revolutionary change, is no longer relevant, and that today it lacks any kind of realistic leverage over power. In addition, what remains is a commitment to the sovereignty of the individual, including, above all, the capacity of individuals to act collectively to achieve rational and nondeterministic ends. More significantly, perhaps, Ambasz's writings and work evince a fundamental respect for the feedback principle in life as in art, in culture as in nature. There is more than a hint in his theoretical position that the principle of homeostasis is much more than just another design option. It is perhaps instead the only liberative design precept that is capable of transcending the current impasse between the reality of climate change and our contemporary propensity for political and ethical paralysis.

The so-called green movement is a big umbrella where, at present, I wouldn't dare to cast too much light because the shadows are still looking for their bodies. It is a state of awareness; it doesn't yet constitute a conceptual reality, because it lacks a precise system of discourse and a theoretical structure that will allow it to transmit a body of knowledge, and to constantly reevaluate it. It is an attitude, so far. EA

REPLIES
to Michael Sorkin's Questions

MS In your lectures you've spoken about the difference between Ambasz and Emilio. What is the difference?
EA Everyone knows that two entities live within me: Emilio and Ambasz. Emilio represents the visionary architect, and Ambasz the pragmatic industrial designer.

Ambasz is an anxious man because he wants his products to be well received by men. Emilio is an anguished person because he hopes, through his architecture, to be welcomed by angels. Ambasz is a sociably sad man; Emilio is a solitary, cheerful one. On the other hand, I am cheerfully sad and sociably solitary.

I was once asked — by you, as a matter of fact — how I would like to be remembered in a hundred years' time: as a designer, a minimal artist, a farmer, a philosopher, or an architect? My answer is: as a poet.

What influence did your Argentine upbringing have on your work?
I was born in Chaco, a subtropical province of Argentina, almost 1,000 miles north of Buenos Aires. Its never-failing afternoon rain could be used to adjust one's watch to 4 pm. Clouds of vapor, evaporating half an hour later, stood as metaphors for the impermanence of all things.

When I was seven years old my parents moved to Buenos Aires. My room on the second floor of a new house opened directly onto the leafy branches of a street tree. With my bed placed against the window it was as if I lived in a tree house. I used to stay up late, in the darkness of my room, looking at the reflection of the streetlights on the tree. I never ceased to marvel at the brightness of a raindrop as it held on to a leaf. I still remember shivering as if caressed by celestial fingers when the rustling leaves made their music. I was entranced by that tree. To this day I revere its brethren.

The stars of Buenos Aires: there are so many more visible in the Southern Hemisphere. Standing on a deep balcony projected over the sidewalk I felt they cast a dome whose perimeter was nowhere but its center was everywhere, while I was nothing. One felt so — but so! — lonely, in such an overbearing universe.

Buenos Aires has always been seen by the Argentines as the incarnation of everything the provinces thought ideal. We knew it was to be a disappointment, but we cherished its pretense to perfection. I have sung to that flawed Buenos Aires, as one would to a beloved son who did not live up to expectations, in an essay I wrote many years ago entitled "Anthology for a Spatial Buenos Aires." Its words and meanings still ring true, only the plaintive sounds of their echoes have become a companion-like murmur.

The pampas, you ask? I remember describing the pampa as an Indian voice for space; land where man stands alone as an abstract being who would have to recommence the history of the species — or to conclude it. I see the pampa as a terse thread, shifting from green to earthy browns knotted onto the Atlantic to the East and onto the Andes to the West. A place of the mind, a melancholy green-gray lair inhabited by a very hard-to-fight resignation.

You have mentioned an Argentine architect who influenced you greatly in your youth and about whom you plan to write. Who was he? What was the nature of the influence?
When I was fifteen years old and still in high school, I came to the realization that one can only learn the craft of poetry from poets. Accordingly, I developed a plan to work for Amancio Williams. A friend who was already in Amancio's architectural office spoke to him about me. I came to interview with Mr. Williams one afternoon — by that time I was already all of sixteen years — and he invited me to join his studio. In order to better attend to this new responsibility, I switched my attendance in high school to night classes and went to Amancio's studio during the day.

Many beautiful things have been said about Amancio's work by Le Corbusier, Max Bill, Mies van der Rohe, and other great artists. I have recently read those over. I am not surprised they had always been fascinated by Amancio's work. To them, he was like Argentina: a child of Europe. Like Argentina, he was the one they envisioned enacting the European utopian dreams. What had been dreamt in Europe was to be given strong poetic embodiment in a virgin place, where memories could only be recalled in libraries. It was the cleanest piece of paper and the largest unspoiled natural surface left. So far away; a place Europe could call its mental dwelling of last resort.

As I look at the panorama of the twentieth century's Latin American art and architecture, Amancio shines as one of its greatest artists. He strongly practiced his belief that architecture must contribute to human happiness; that for architects to revel in historical and/or simpleminded methodological references was to skirt their responsibility.

He always believed in creating and inventing master examples. He constantly searched for the irreducible solution, believing that if an architectural problem could be reduced to its essentials its answer would stun evil and proclaim God's kingdom. He searched for prototypical pilot concepts. They were to be as unselfconsciously simple as many a child's answers. After all, what could be more obvious than to put Buenos Aires' airport on the water of the River Plate, where it would not have created urban and traffic conflicts, where it could have been easily erected by barges effortlessly bringing materials to the site, where it would stand out in poetic contrast with the river's long brown horizon that Amancio's geometrical planes were to stitch to the blue sky?

It must be said in his defense that Amancio never spoke in romantic terms; he always explained his projects with a sometimes overbearing abundance of technical details. These were always right, even in their most extreme cases. I have always suspected that by analyzing the pragmatics of his projects to such an exhaustive degree Amancio made it possible for his explanations to become an intellectual superstructure, a shell under which dwelt the poetic core. Perhaps Amancio believed that such a blanket of technical perfection would everlastingly shield *"Teknes'"* marriage to *"Poesis."*

Let these heartfelt words stand as my testimonial for all the magnificent images he has created. If someone has called him Classical—perhaps a misnomer—it is because his work is so essential, so irreducible, and so luminously strong that it transcends materials and construction methods to embody the spirit of architecture. The country was created, it would seem, yesterday, but it is only when great artists like Amancio appear that we are able to evoke a notion of dwelling in peace with ourselves. I do not know a greater accomplishment for an architect than to have created such magnificent abodes of the heart that we can find refuge and solace even when we are away from them.

Why do you divide your time between New York and Italy? What are the practicalities? The embodied meanings? Which does Emilio prefer? Which Ambasz?
If Emilio could not reenter Italy he would feel as he had been thrown out of Paradise. If Ambasz could not return to New York he would feel excluded from what was the capital of the twentieth century. As for me, I need both places: my feet on the garden's earth, and my head in those constantly changing clouds, which are suggestive of things to come.

Will Eden keep safe its magic; will New York maintain its lead in this new century? Maybe I am positioned in that borderline moment when a fruit is about to turn ripe: its maturing attractions still strong, its decay barely suggested.

As for my practice, I could work perfectly well anywhere— Patagonia, or the Fijian islands. Email has set me free of a required local. Personal relations still count, indeed, and I maintain these by meeting regularly with my collaborators.

Certainly Japan and Europe have been more generous than America regarding work opportunities. It has taken me thirty-five years to prove to them the practical advantages of my ideas. I am more than willing to wait another thirty years for a call from the USA. In the meantime, I have begotten children, grandchildren, and not a few little bastards. To see Renzo Piano, Jean Nouvel, Tadao Ando, and many other excellent architects, utilize vegetal matter in their projects makes me feel my mission is beginning to bear fruits. To hear some of them claim these ideas' paternity makes me feel like a mythological character, but I know it is just a case of a foretold Freudian destiny.

How do you see your work in relation to the global environmental movement? Does architecture matter here?
Architects have always bled for the ills of the world. In my view, they will stand far back in the line of those sent to Hell for their environmental sins. But there they will go, inevitably, if they do not honor their ethical responsibility to propose alternative models of the future.

We must create alternative images of a better life to guide our actions, if we do not wish to perpetuate present conditions. I believe that any architectural project that does not attempt to propose new, or better, modes of existence is unethical. This task may stagger the imagination and paralyze hope, but we cannot subtract ourselves from its pursuit.

As a ancient partisan of class warfare, whose side do you think you're on?
I've never been so much a man of leanings as much as I have always been a man concerned with going forward. I have always striven in my work to present alternative models of the future so we can change the present. This is a task to which I have dedicated my heart and mind.

My existential wager is on poetry, and a commitment to justice is, for me, one of the necessary but not sufficient conditions to achieve such a high plateau.

Justice, whether social or moral, is a conceit of the mind. Justice does not exist in Nature, but despite this cruel fact, I feel very strongly that it is our ethical imperative to pursue its implementation on Earth. Even if we know it to be a delicate structure, held together by such ethereal material as abnegation and altruism but destined to collapse at night, we must rebuild it every morning.

Why aren't your industrial design objects fuzzier, like the buildings? Describe continuities and discontinuities between the two practices.
By designing products, man extends his grasp. Directly related to his body, a product answers to the parameters his anatomy imposes. Creating buildings, we seek to make these in the image of a compassionate microuniverse.

Products are prospedeutic; they represent prosthetic extensions of man. They stand as surrogates for those capabilities man is short of, or lacks completely. A building is, in a certain sense, also a prospedeutic conceit but it is also, and more importantly, a representation of a harmonious place. A place where man is protected and provided for, where he is able to exist as himself: a lair where he might even become at peace with himself and the universe outside.

If we were welcome on Earth we would only need to lift our hand and catch a banana, or put a finger in the ground and get a telephone connection. We have been told by all legends that we have been cast on Earth to toil and suffer. We are of this Earth, but she is an indifferent mother; we are neither fully adapted to nor adopted by her. Without going back to Adam and Eve it is true that we are not fully welcome here, that we have to constantly construct peace treaties with Earth that we know will be temporary. For me, a house is a formal manifestation of such a pact of mutual respect. I understand the house as an abode, from the verb "to abide" — what the French call *demeure,* the Italians call *dimora,* and the Spaniards call *demora* — a place where we dwell in the span of time allocated to us.

Going back to your question from a different angle, let us look at the case of the Vertebra Chair. Vertebra is an office chair that I codesigned and developed, which was presented in June 1976. It was the first articulated chair designed to respond automatically to the movements of its user. It was a forerunner of the high-tech look, a highly mechanized artifact that you might call a nonfuzzy object. I was very concerned with creating anthropomorphic and anthropofunctional objects that would accompany the body's movements completely and unselfconsciously, just as a glove moves with the hand wearing it. The chair was conceived as an extension of the body; this was the underlying principle guiding its design. The form and stylistic appearance I originally gave to the chair would very likely be different today; look at my more recent office chair designs, which are still faithful to the original principle, but have now been given a soft-tech expression. This has come about because my stylistic interests have undergone changes, but the essential concept behind the articulated chair remains the same.

It seems that you have a problem reconciling the stylistic guise of Vertebra with that of some of my architectural projects. Please remember that while men sit on chairs, and their bodies determine a great deal of the products' form, they dwell in buildings at some distance from the walls. The further the artifact moves away from the user's body into his social and psychological milieu, the more architectural design becomes determined not only by ergonomic considerations, but, more strongly, by its surrounding physical and sociocultural domains.

Do you believe in the Gaia theory?
Gaia is to me not a theory, but a poetic hypothesis, and I am all for poetry. Although it is a metaphor, it provides a temporary structure on which to affix many seemingly unrelated questions. When seen from a certain distance, that frame of reference may help us to connect some of those disparate points into a cohesive theoretical structure.

Place yourself within the context of current architectural production. Jim Wines? Ken Yeang? Michael Reynolds?
Many years ago, when Peter Eisenman got me promoted from freshman to senior status during my first semester at Princeton, he told me, standing in one of those academic corridors where eunuchs dream of power, that now I was a senior he would make me into his "model" student. My mouth opened, and, as if governed by a spirit somewhere in the pit of my stomach, I heard myself saying, "I cannot be your model, Peter, because I'm going to be the father of a nation." I was as astonished by my words as Peter was; for once in his life, he was speechless.

I know it sounds presumptuous, but I lay claim to being the precursor of current architectural production concerned with environmental problems. If there is any strength to my architectural ideas, it comes from the fact that I believe architecture must be not only pragmatic but must also move the heart.

I rejoice immensely when I come upon somebody else's work that touches me, even if it is the architecture of someone such as Gehry, whose work is so different from mine, and whose concerns are totally unrelated to mine. What matters to me is that he sings his own song. His birds may not alight often in my garden, but I'm sure they will pollinate even my own flowers.

As for those who practice my architectural credo, I am not interested whether they cover their work with salad, but whether their work can strike an emotional chord. In that regard, I'm reminded of when I was very young and I created in Buenos Aires a group named Friends of Contemporary Music. With other friends we hired a small auditorium, and ran around looking for people who wouldn't mind studying or knitting while listening to this type of music, because I was ashamed we were so few and the musicians had worked so hard. After listening to Xenakis, Stockhausen, Davidovsky, Kagel, and many others I came to the realization that the idea was there, but that it did not move me; the poet of electronic music had yet to arrive. I am eagerly looking for such a person in my field of endeavor, and if I don't see him yet it is not, I hope, due to blindness, but perhaps because I am biased.

Unlike many, you have managed to successfully skirt involvement with the academy. How? Why?
I have not completely skirted the academy. I was a teacher at Princeton University's School of Architecture. The same week I was made the Phillip Freneau Preceptor of Architecture, I resigned. The dean of the faculty was livid. "One does not resign," he hissed, "when one is appointed to such a position at twenty-five years of age. You are very tired and what you should be asking is for a one-year leave of absence." I replied that I wasn't so much tired as very concerned that I was misleading my students. I felt very strongly that architectural education as taught was no more than a somewhat stylized simulation of office practice. I was utterly unconvinced of the teaching methods utilized. But I was even more certain that one needed to create a new type of *Universitas:* that is to say, a new type of institution concerned with the design and the management of the man-made milieu. This was the task I set for myself, and I was resigning from my position so I could pursue it. This idée fixe gave later birth to the *Universitas* project I carried out as the curator of the Architecture and Design Department at the Museum of Modern Art, New York. But this is a subject for another book.

Does contemporary architecture have a universal subject? Or may I rephrase this: Does your architecture have a universal subject?
I have always believed that architecture is an act of the mythmaking imagination. I believe that the real task of

architecture only begins once functional and behavioral needs have been satisfied. It is not hunger, but love and fear — and sometimes wonder — which make us create. The architect's cultural and social contexts change constantly, but his task, I believe, always remains the same: to give poetic form to the pragmatic.

There is in all of us a deep need for ritual, for ceremony and procession, magical garments and gestures. It is an archetypal quest in which we all partake.

In my architecture I am interested in the rituals and ceremonies of the twenty-four-hour day. I am not interested in the rituals for the very long voyage — a voyage which can take forty or fifty years. And what a tragedy to discover that, for the sake of those long-term dreams, we have sacrificed our daily lives. No, I am interested in daily rituals: like the ritual of sitting in a courtyard slightly protected from both the view of your neighbors and the wind — looking up to the stars. Dealing architecturally with that type of situation attracts me. Because the magic of daily existence is not in the house; the house only provides a backdrop for it.

The architecture I create is steeped in mysticism. On the one hand, I am playing with the pragmatic elements that come from my time, such as technology. On the other hand, I am proposing a certain mode of existence which is an alternative, a new one. My work is a search for giving architectural form to primal things: being born, being in love, and dying. They have to do with existence on an emotional, passionate, and essential level. Perhaps I use very austere elements to express this quest and, therefore, the gesture may be seen as an austere one also. But by doing it in this way, I believe that it may be far more durable — it is certainly a far more classical attempt. I am interested in the passionate and the emotional when they assume a timeless guise.

In one of my projects, the "House in Seville," I actually wanted to eliminate architecture. The only thing to stand was the facade, which would be like a mask — a surrogate for architecture. The architecture would disappear. You would see only the earth. You might say that by this device I rhetorically sought to eliminate architecture as a culturally conditioned process and return to the primeval notion of the abode.

I seek to develop an architectural vocabulary outside the canonical tradition of architecture. It is an architecture that is both here and not here. With it I hope to place the user in a new state of existence, a celebration of human majesty, thought, and sensation. Though apparently quite new, there are devices — both primitive and ancient — permeating the designs. The result is an architecture that seems to stand for eternity.

The ideal gesture would be to arrive at a plot of land so immensely fertile and welcoming that, slowly, the land would assume a shape, providing us with an abode. And within this abode — being such a magic space — it would never rain, nor would there ever be inclement phenomena of any other sort. We must build our house on earth only because we are not welcome on the land. Every act of construction is a defiance of Nature. In a perfect Nature, we would not need houses.

If one finds the quintessence of a problem, one will have better access to an irreducible solution. The thread supporting my design quest in every area — my products and my architecture — is a single preoccupation: finding the root of the problem, its essence.

As for expressive means, I seek to approach a design problem in the most crystalline, austere, and graceful manner. I long for an architecture which has been reduced to essentials and which, at the same time, is an architecture full of potential meanings. Such concision is the method by which to achieve a multidimensional, epigrammatic architecture.

If I may paraphrase Paul Valéry: my quest for the essential in architecture is not about being simple and light like a feather; it is about being essential and concise, like a bird.

Architecture is, for me, one aspect of our quest for cosmological models. I suspect that such an all-encompassing image, if it ever arrives, will be as simple and as dense as a point suspended in midspace. Every one of my projects seeks to possess at least an attribute of the universe. The quest for that which is infinite, eternal, and ever-present, I suspect, may be contained in designs of very few lines which manifest themselves with great economy of expression. In such a seemingly simple manner those lines may, hopefully, acquire the fascinating power of mythical structures. Maybe it is because I seek essentials that I love Lucretius's *De rerum natura* so much.

I am only interested in discovery, not in recovery; in invention, not in classification. In the uncharted realm of invention, taxonomy is a process yet to be born. In the same way, as I search for essential and lasting principles in architecture, I have been told — by you, as a matter of fact — that in opting to write fables rather than theoretical essays I have grasped something basic: fables remain immutable long after theories have crumbled. The invention of fables is central to my working methods and not just a literary accessory. The subtext of a fable, after all, is a ritual and it is to the support of rituals that most of my work addresses itself.

What have you been reading lately? Is there anyone from the architectural-theoretical wing thinking along lines sympathetic to you?
I have again been rereading *Morceaux choisis,* Paul Valéry's own selection of some of his best essays. I have also just finished reading Susan Sontag's *Where the Stress Falls.* While Valéry always elates me, Sontag needs someone to let the air out of her balloon. I am sadly disappointed by most of her essays' lack of intellectual rigor. But perhaps I'm wrong — maybe I am dense to the altruistic fact she's just trying to emulate Pavarotti; you first sing to the masses a bolero, then a mambo, and slowly you get them to *Wozzeck.* Maybe that's the way she generously planned to introduce people to literature and philosophic thought. But by stooping so much, she ran the risk of remaining bent.

To think that I discovered her in 1963, during my short interlude as a freshman! I read an article she wrote for *Sight and Sound,* a British film magazine, on Robert Bresson. I thought it was extremely subtle and perceptive. I brought it to the attention of Professor Szathmary, director of Princeton's Creative Arts Program. He made funds available for me to invite her to Princeton to give a lecture. When she came off the bus, the first question she put to me was, "Why was I invited?" It was the first time, she told me, that a university had done so. I am afraid that life tends to turn such youthful sincerity opaque. In her early days she radiated wonder. Let me keep that memory of her.

You asked about my recent readings of architectural-theoretical thinking. I am intermittently embarrassed and irritated by the woolly-mindedness of some of the writings on architectural subjects with which I am sympathetic. No doubt I am unlucky in my choices. There must be pearls out there. Would you consider sending me a bibliography?

Your work is vividly pastoral. Do you have any hope for the future of urbanism?
Where did you get the idea that pastoral fields existed only outside the city walls? To this day, you can go up to the top of the towers of a medieval city like Bologna, and discover that behind the facades defining treeless streets exist immense gardens which occupy almost 25 percent of the city area. Those were once vegetable gardens, and places where cows grazed. Those grounds were of utmost importance to survive a siege.

The well-intentioned Green City movement was quite illiterate in terms of history, like many other endeavors single-mindedly committed to "progress." I strive for an urban future where you can open your door and walk out directly into a garden, regardless of how high your apartment may be. I submit it that my building in Fukuoka is one example of how we can, within a high-density city, reconcile our need for building shelters with our emotional requirement for green spaces.

What is your taste in cinema? Which films do you particularly admire?
My favorite film director is Robert Bresson, followed by Friedrich Murnau; and I would be an ingrate if I forgot Fellini. As for actors, I rely on the director to turn even a cow into a delicate dancer.

One of your projects is called a "real-world theme park." Is this an act of resistance to the Disneyfication of the planet? Would you care to dilate on what seems to be the theme parking of everything, including the profession of urbanism?
"Real World Theme Park" is a project I did at the request of Peter Gabriel. Those were the years when he flew in on the Concorde to discuss his ideas in JFK's VIP lounge. After explaining his goals, he flew away that same day, again on a Concorde.

You ask why people wish to adopt the behavioral strictures that a theme park provides. It is because we have become bereft of rituals and ceremonies. We do not know any longer to whom to pray, and are encouraged by the real masters to delude ourselves into thinking we are free to shape our destiny. We have been brought up to believe we can become anything we want. The problem is that one can only want something new if one can imagine it. If we can't do so, we can only fancy ourselves actors playing diverse roles in a theme park's prefabricated script. We can stupefy ourselves by switching roles, no longer remembering who we were when we started. Theme parks are strong drugs that allow dreaming without the risk of mental alteration. If, by doing so, anxieties can be anaesthetized, why do you wonder that people crave such seemingly inoffensive distractions?

You ask what relevance the concept of "theme parks" may have for urbanism. If it is now possible to live anywhere in the countryside, cloud deep in the belief that one is connected with everything and anyone via the Internet, how can an urban planner refer to such an elusive and amorphous sociocultural system as a guide to creating urban forms? Unless he is stiffened by strong ethical beliefs, he will opportunistically offer to design the palaces and temples of the emerging powers and their manipulated beliefs. That is the goal this season's prophets propose as urbanism's task: to return to its old profession of glorifying the masters and mystifying reality. Aren't the siren songs that invite architects and urbanists to remain the Present's servants what are most praised by academia and relentlessly taught in its schools?

What is the deal with those trees up in the air?
If you're not referring to some beautiful drawings on this subject that Max Peintner drew more than thirty-five years ago, I must confess I am a bit lost.

Do you perhaps refer to the project in Japan, where we cut a large hole in the roof of a parking garage, and suspended there, by means of a metal truss, a tree with its trunk midway, its roots under the roof, and the branches above? To avoid people getting caught in the interspace between the trunk and the hole's edge we put up a transparent parapet. This parapet also helped to keep contained the misty cloud of water used to water the plant.

This tree grew to the point that its expanded trunk closed up the hole. How could this come about? The tree's roots drew savory nourishment from the carbon dioxide in the exhaust fumes of the cars.

What is the difference between ornamentation and camouflage?
Camouflage is used to subtract, or to hide something already existing by giving it another guise: for example, covering a war tank with a green spotted tarpaulin to transform it into a part of the landscape. It is a conservative act. By dressing it up as something else it remains as it is.

Ornament is used to assign something the attributes of something else we want it to become: for example, when the capitals of columns and the cornices of buildings are dressed up with stone-made flora. The key word here is "attribute." It is an act requiring the suspension of disbelief. We know these elements not to be what they represent, but we accept them as if they were endowed with the characteristics with which they present themselves. It is an act whereby we invoke a reality not truly present.

Using elements from the vegetal kingdom I do not introduce ornament but function. Not only is the introduced element exactly what it purports to be, but it is intrinsically related to the building's proper performance. I grant you that it might be perceived as an ornamental process. However, to me, that denomination—although mistaken—reveals a desire to satisfy an unrequited longing for adornment, a yearning the Modern movement has neglected. I presume to satisfy it by using true elements vitally rooted into the building's matter.

I am inescapably a child of my time, and therefore distrustful of making irretrievable formal statements; thus I use elements that change with the seasons. The leaves fall in autumn allowing the sun to warm the walls, after shading them from the heat in the summer. It is an adornment, indeed, but it is intrinsic to the building, just like breathing skin is to an organism. When marble leaves and garlands fall to the ground or end up in museums, the building is scarred but its usefulness survives. If the plants covering my buildings are torn off, the building suffers a substantial reduction of its raison d'être. If these plants had been just an ornament, removing them from the building should detract from its visual characteristics, but not from its integrity and performance.

Could it be that mimesis is an answer closer to your original question? Perhaps, but only if we keep this discussion in the domain of rhetoric. I must insist such is not my case. The vegetation used does not try to assume the shape and the attributes of an already existing entity, as marble leaves do. It is itself, not a representation of it. By using these plants and moving the earth around the building I seek to reconcile the building's existence to itself as a principality and to its larger context. To me *reconciliation* is the word you are searching for.

Let's go back to the beginning. If Nature were to have welcomed us in the guise we have become, we would have needed to make no shelter. Do you believe that I seek to lift up the corner of a grassy meadow to cover myself as if with a magic blanket? That by so doing I try to domesticate Nature? I believe that in our pursuit to master Nature-as-found, we have created a second man-made-Nature, intricately related to the given-Nature. We need to redefine architecture as one aspect of our man-made nature, but to do so we need first to redefine the contemporary meaning of Nature. Perhaps a new philosophic academy is what is called for. Shall we call such an institution a "Universitas," that is, the whole?

You have long been interested in large-scale forms of organization and production. Would you address this briefly, particularly as regards the role of the architect. (We know, for example, that Rem and the Dutch promote the metaphor of the architect as "surfer," riding the wave of globalization.)

Short Answer:
S, MS, VS, and XS (small, medium small, very small, and extra small).

Long Answer:
Convinced that architecture should be the work of someone else's imagination, some celebrated architectural contributors to a marketplace culture repropose good old Neo-Modernism 101—tilted, twisted, and fragmented; oversqueezed and left for dead, but still juicy—as the easily disposable clothes to dress à la mode chapels for those believers in "I consume, ergo sum."

Do you believe in the sublime? (Or a sublime?)
If an architectural work does not move the heart what is the point of it? It is just one more building. I am not ashamed to say I pursue the transcendental in my work. It is ungraspable and elusive, but I am devotedly convinced it can be evoked.

Frank Gehry is always on about the influence of artists on his work. How do you see yourself in relation to contemporaneous trends in the art world: to Heizer, Judd, Serra, Miss, etc. Are these relations important? Do they ever flow two ways?
Robert Smithson and I instantly became great friends. At the end of our first dinner together, we felt like long-lost brothers. The same things interested us, from different viewpoints. I was fascinated by his work, and I presume he was interested in mine. Never again did I have such a feeling of conceptual brotherhood with anyone. I adored my conversations with him, and I was devastated when he died.

I have also maintained a very good friendship with Sol LeWitt, whom I have always considered the guardian angel of the Minimalist group. Michael Heizer approached me once with a request that I write the introduction and a critical essay for a book on his work. I knew his work, of course, but I had never met him before. I presumed he was interested in the fact that we shared many formal interests and affinities.

It may come as a surprise to you, but I was not acquainted with the work of Richard Serra when I designed the Mexican Computer Center in 1975. The first time I encountered his work was at the Yonkers [Hudson River] Museum. During the installation period, I was taken to see the exhibition by the museum's director, Richard Koshalek. I was dazzled. I felt that if I ever made a sculpture I would be very happy if it looked half as good as Serra's.

Those are the bare biographical facts. As for influences, I think that you are a much better judge of how they flew, and how they flowered.

REPLIES
to James Wines's Questions

JW We have both pursued an environmental sensibility for many years now and, even though a response to context has defined virtually all great architecture of the past, the mainstream design scene has continued to marginalize our work. To what do you attribute this peculiar situation?
EA Anyone who seeks to speak in a new language should not be surprised if he or she is not understood. They will certainly be kept on the margins. And, frankly, I'm not surprised that this happens — and in fact I would be greatly concerned if I got a plumber replying: I agree, pal.

Perhaps my disregard for plumbers' approval originates in the fact I do not care for the honor that men may bestow, but for the smile of angels.

I divide architects into three categories: first, the commercial architect, what some people would call the "hack" architect; second, the professional architect who practices with a high degree of professionalism within an accepted discourse; and, third, the inventive architect. These three categories correspond closely to the notions of stereotype, type, and prototype. He who invents a prototype is fracturing artistic convention, extending present lore, and introducing new symbols not immediately understood. Once a culture starts understanding the meanings embodied in those prototypical symbols, the prototype becomes an understood type. As the culture that once gave context to the type starts shifting away, the type becomes a stereotype. Those who dwell in a culture that is homogenized and pasteurized feel more comfortable with stereotypes. This is a possible explanation as to the why of my serenity when I am marginalized. Sometimes, when I feel too lonely, I fancy that I am on the edge of the forefront.

Years ago — it has now changed — whenever I gave a lecture I felt a wave of hate coming from the audience of architectural professionals. Maybe it was because I was no longer only presenting ideas but I also built buildings. They could no longer dismiss my images as "easy to draw but let's see if he can build them," because they were already built. In essence, the buildings were too strong a demonstration that they had put an existential wager on an exhausted horse. Some of them in those years gone called me "the Antichrist." Now, some call me the Messiah. Tomorrow, what?

With every concerned environmentalist and sociologist in the world telling us that vegetation, urban agriculture, garden spaces and forestation are essential components of the city (for reasons of health, well-being, and psychological stability), **why do you think the mainstream design world still resists the use of landscape — or, at best, sees it as some kind of peripheral décor?**
As you well know, he-men architects look down patronizingly on interior architects and exterior ones — meaning landscape architects. They feel very strongly that theirs is a true embodiment of the normal and natural, and that those other two categories are at best craftsmen or hairdressers. The great majority of architects who graduated from academia have been given the fortitude of a systematic métier, and the conviction that they belong in this world, if not outright own it. They have been taught that little square windows, or sliced, protruding, exploded, spiraled, twisted, and tilting planes — some of them in ways that Escher would appreciate, depending on their schools — are architecture. Their professors rewarded them handsomely for toeing the party line. How can you expect them to utilize materials that are not the traditional ones; how can you expect them to try to integrate a building with Nature when they are the proud heirs of a Greco-Roman tradition of mastering Nature, standing above and distinct from it?

Although the architects bleed for all the sins of the world with their mouths, many of them will not dare to puzzle their clients with an architecture that is humble and a part of Nature, for fear of losing them. Actions and ideas like those that you and I propose are quixotic and reckless; better practiced by those who have no families to support, or who fancy that barefoot children grow up healthier.

How has your thinking about architecture and its role in social and environmental reform changed in the past few years? Are you still as idealistic and as hopeful as you were twenty years ago?
For me, the definition of courage is not that of someone who marches into battle unconcerned, but that of someone who is trembling, yet nevertheless marches ahead because that is what he must do. The lucidity of fear, if it doesn't paralyze, is a badge of honor.

I always knew that my pursuit of alternative models for a better future would be rejected and mocked, or, at best, that I would be left alone to bark at the Moon. But I always remembered that although the madman who threw stones at the Moon never hit her, in the end no one else in the village could throw as high. I still feel idealistic. I cannot say that I am hopeful because by nature I'm a pessimist. Whenever I go forward and fight on, I also have to fight within myself a pervasive awareness of the vanity and pointlessness of it all. But I devotedly believe that, although the task of inventing better futures may stagger the imagination and paralyze hope, we cannot relinquish this holy call.

Clearly, now that we have a little perspective on art of the twentieth century, those artists who dealt with psychology, ambiguity, morphology, context, chaos, and other indeterminate elements affecting people's relation to their environment are rising as the most interesting, relevant and enduring artists. I am thinking particularly of such non-formalist practitioners as Duchamp, Johns, Rauschenberg, Bacon, Giacometti, Smithson, and Matta-Clark, and writers like Beckett, Artaud, and Burroughs. So why has architecture remained so relentlessly and dogmatically locked into the lessons of abstract art, when this entire tradition seems to be growing less interesting with time?
I've always been aware, and so have you surely, that architecture has constantly marched behind the other visual arts. Think back in time and you'll find that painters anticipated many of the concerns that later were taken up by architects. Architecture is a social art, and the human material it utilizes is far more recalcitrant than oil paint.

You mentioned nonformalist practitioners, but I assure you that the same holds true for formalists such as Gris, Braque, and, of course, Picasso in his Cubist period.

You ask why architecture remains locked into the dogmas of abstract art. Just think of your students. Which of them dares to make a curvilinear gesture? Even people in my office only feel comfortable when I make it, provided they can reconstruct such a line when they discover its loci (or assign it one) with a compass in hand. It was useless for me to break their triangles and T-squares; the computer brought these back through the big door. There is great fear in making the graceful line without the directorate of a ruler. I'm sure you have observed that architecture students are more frightened of their own shadows than painting students are of real phantoms.

Maybe that is why we walk outside the established path; you were an art major, and I never got an architectural education, on account of my having completed undergraduate and graduate work at Princeton in two years, never having taken courses in plumbing or the practice of the profession, but in philosophy, literature, and sociology.

You often speak of your work as mystical, cosmological, and concerned with myth. In a world of "simulacra," where culture and spiritual values are being abandoned and/or purposefully destroyed, where do you find your audience?
Here I cannot help but give you an answer that will inevitably sound arrogant: I let them come to me. I make the image; if it grips their heart they will listen. For me, an audience of four in a multitude who looks the other way is enough. Those four have been touched, and that's the only audience I care about. I have trained myself since I was a teenager, and observed the reaction of my classmates at Princeton when I presented my projects, to just aim to please myself. If by so doing I attract an auditor, so be it, as long as it is a critical auditor, one of the inquisitive sorts. I have no patience for ungifted admirers.

Where do you see the green movement going in the future?
It will branch out. The so-called green movement is a big umbrella under which, at present, I wouldn't dare to cast too much light because the shadows are still looking for their bodies. It is a state of awareness; it doesn't yet constitute a conceptual reality because it lacks a precise system of discourse and a theoretical structure that will allow it to transmit a body of knowledge, and to constantly reevaluate it. It is an attitude, so far. It is not yet a principle. Do we want to let the few acolytes we have know it? Reckless is the shaman who shows how his magic is held together by stitches.

Are you interested in recent developments in environmental technology and are you using some of these innovations in recent projects?
I am very much interested in all sorts of technology. As a matter of fact, I am one of the few architects who not only designs his work in 1:1 scale detail, but also knows how to mass-produce these details. Obviously, this is the result of my also being an industrial designer who designs, engineers, and solves all sorts of production problems presented by the industrial products he invents.

I believe that the only way to actually solve the problems that technology may impose upon society is by using technology. The problem with a technological society is that it is technically illiterate, and therefore under the spell of techniques, and the rule of its high priests. But one should not confuse the use of techniques with architecture. I don't believe that the pyrotechnic use of technique is enough to endow a building with an architectural spirit. Without poetry it is no more than an exodermic skeleton.

Which artists, architects, and writers from the past (before the twentieth century) do you consider as your influences? Which do you most admire?
Now, here you've got me. Of the architects I feel that I have been very influenced by I must mention, first of all, King Solomon and his temple in Jerusalem, although I have yet to see a color photograph of it — even a B&W would do. Going back a little bit further, I am a devotee of Imhotep. Then I jump ahead to Palladio because I have a weakness for practical and pragmatic architects. Based on the aforesaid, it shouldn't surprise you that I admire Jefferson, but my heart, or whatever I use as its surrogate, belongs to Frank Lloyd Wright. Now, since Terry Riley, former director of the Museum of Modern Art, New York, Department of Architecture and Design, has called me, in print, "the greatest architect of the enlightenment," do I qualify, at least in a temporal sense, to be included among those I might select from?

The truth is that I have a debt of gratitude to Bernard Rudofsky's book from 1964, *Architecture Without Architects;* an introduction to nonpedigreed or N. N. architecture. I could list many N. N.'s who, without any diploma, rank high in the firmament of my greatly admired architects.

As for musicians, in the domain of operas I start with Monteverdi, jump immediately to Mozart, and then go on to a few, but not too many, arias by Puccini. As for more contemporary composers, I melt when I listen to Debussy,

Satie, Ravel, and that bombastic but extraordinary Stravinsky. As for writers, I am full of admiration for Lucretius; Ovid is not too far away. As for more contemporary writers, I keep returning to Henry James by walking around Proust.

We have both fared much better with patrons — as opposed to clients — for the sponsorship of our work. Where do you find patronage today, and what is the difference in the climate of such support from the situation twenty years ago?
I would distinguish our patrons or clients by dividing them into private and public clients. My private clients, in 95 percent of the cases, have been other architects who, when presented with a problem requiring a more holistic and environmental approach, came to me for design ideas. As for the public clients, they are also architects, in this case younger ones who have become civil servants within City Hall and, when given an opportunity, look for me. So I have touched the hearts of a few architects, after all.

Which of your works do you consider the most important, and why?
Of course, the most important is La Casa de Retiro Espiritual, a.k.a. the "House in Seville." With it I actually wanted to eliminate architecture. The only thing to stand was the facade, which would be like a mask — a surrogate for architecture. The architecture would disappear. You would see only the earth. You might say that by this device I rhetorically sought to eliminate architecture as a culturally conditioned process and return to the primeval notion of the abode. Contrary to everybody's expectations and hopes, it was actually built, and stands proud and handsome.

Another important project for me is the one in Fukuoka, because it demonstrates that you can have the "green and the gray," one on top of the other, and at the same time (to hell with Hegel and his antithesis) you can give back to the community 100 percent of the ground that the building's footprint covers in the form of gardens accessible to everyone, from the ground floor. This building is, for me, very strong evidence that the prevailing notion: "the cities are for the buildings and the outskirts are for the parks" is a mistaken and narrow-minded idea only favorable to commercial architects hell-bent on their not-so-well-rewarded task of enriching developers. The Fukuoka building demonstrates, once and for all, that you can have a building and a garden that are 100 percent of the building needed by the investors and 100 percent of the greenery longed for by the building's users and its neighbors.

On a conceptual basis, what do you consider to be your biggest breakthrough?
I think my biggest contribution is to demonstrate that you can evoke the spirit of architecture without using the canonical elements of traditional architecture; that you can bring a dwelling to life even using bales of hay.

What do you most dislike about mainstream architecture today? And what do you like? (I'm not referring to individual architects here but, rather, tendencies and directions.)
I dislike with all my heart the members of *Das Internationale* of the Square Window (forgive me, Josef Hoffmann; your *Quadratls,* those are indeed full of grace). I have a great amount of disdain and undisguised rage against those who keep on squeezing the Modern movement, now completely understood and stereotyped, by twisting walls and bending corners. To me they're nothing but salesmen of old soap in a new package.

The Museum of Architectonic and Design Arts

Madrid, Spain, 2011–2012

Founded by LEAf, a private Spanish foundation, this museum is to be built in Madrid, on the Paseo del Prado 30, facing the world-renowned Museo Nacional del Prado and Madrid's Botanical Garden. It will become a part of what is becoming known as Madrid's Museum Promenade (with the Museo Nacional del Prado, the Museo Thyssen-Bornemisza, Museo Nacional Centro de Arte Reina Sofía, the Fundación MAPFRE, Casa de América, and the Museo Naval de Madrid). Its program will be dedicated to the notion of architecture and design as art forms. It will celebrate those projects, buildings, and objects that move the heart, proposing alternative models for a sustainable future to guide our actions in the present. The site for the museum is on a prominent corner (Paseo del Prado and Calle del Gobernador) and is L-shaped, with a total surface of 8,000 square feet (743 square meters). The building will have 36,000 square feet (3,344.5 square meters) over five stories plus a basement.

All floors except the top one, which will be a restaurant as well as housing the administrative offices, will be dedicated to exhibition spaces. An auditorium will be situated in the basement, along with the building's mechanical equipment. This museum is designed to meet all the requirements imposed by ICoM (International Council of Museums) if museums are to be granted top ratings.

The building's entrance is defined by two large rectangular and perpendicularly inclined planes, covered with greenery and "leaning" on each other by way of their touching corners. These two planes—essentially the facades—are built of reinforced concrete, as is the building's structure. A sophisticated watering system that also serves as a fire deterrent insures the plants' longevity. Three glassed vertical planes on each of the facades and in the building's center will help visitors to orient themselves on each floor, vis-à-vis the surroundings, as well as enabling them to admire Madrid's Botanical Garden and the Museo Nacional del Prado across the street.

Associate Architects: F. Mariani, Valladares & Assoc.

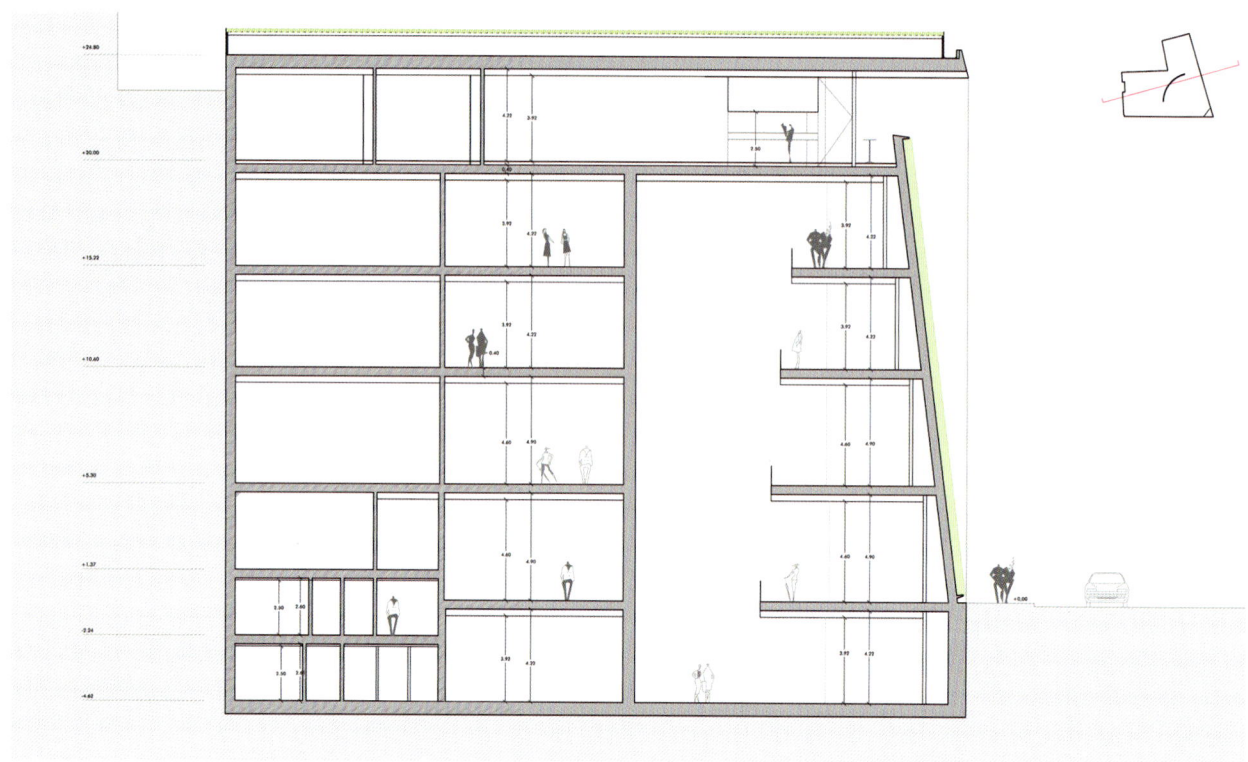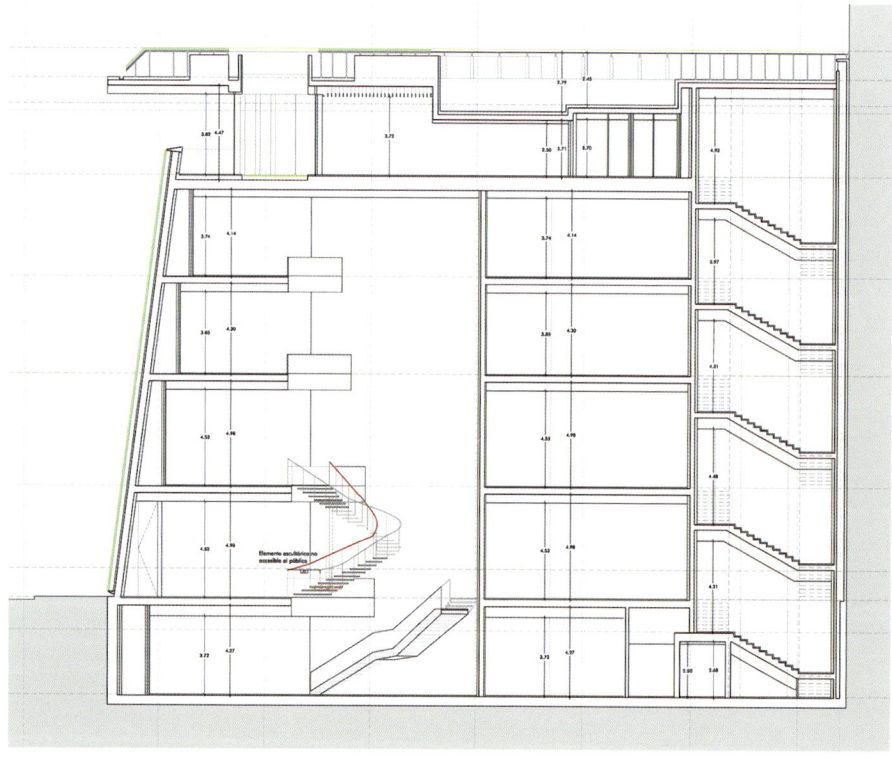

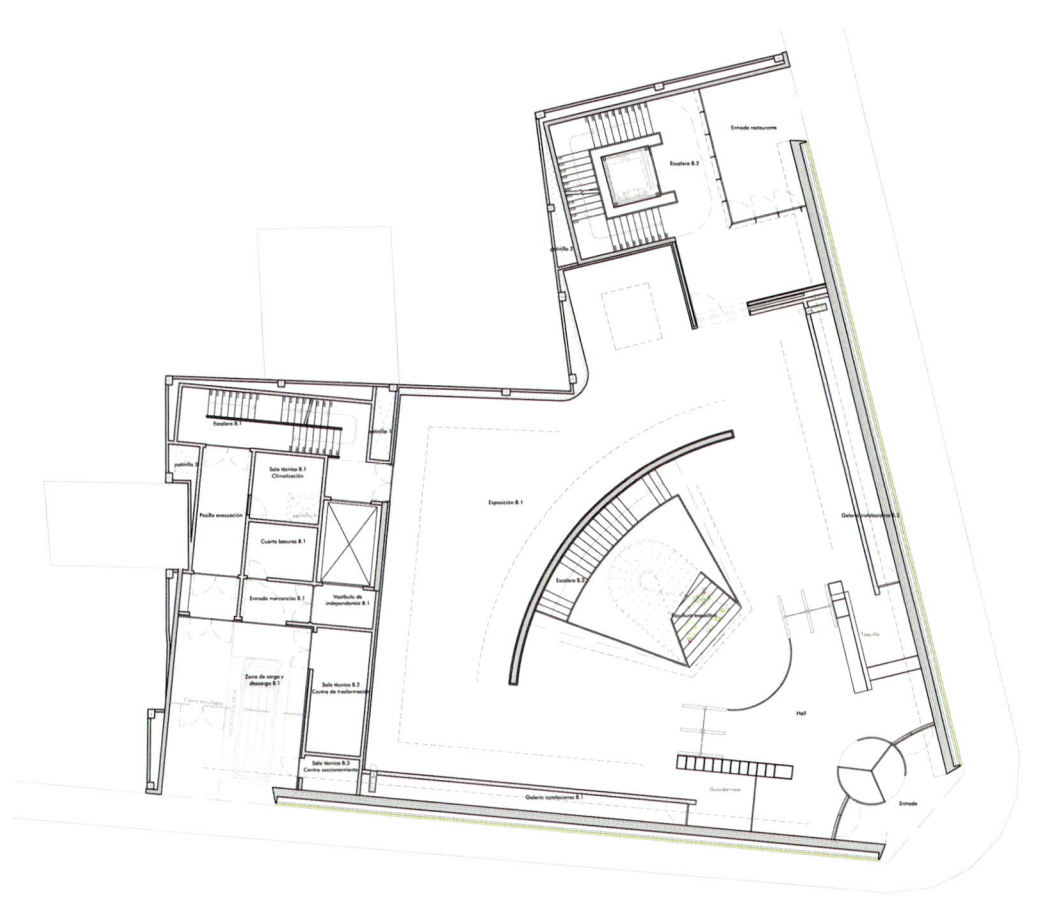
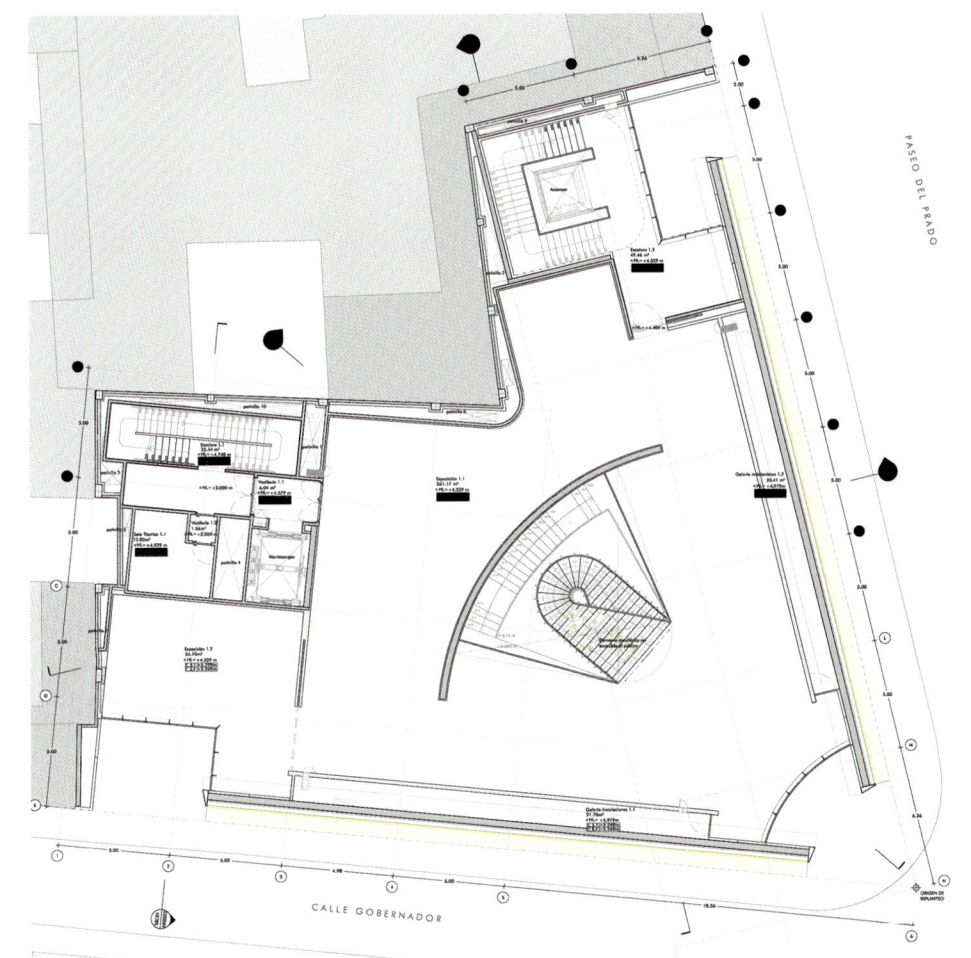

Fabula Rasa

*The little village was in the grip of
fear; fear of Divine rages and of human
passions. One of the men started to
build a construction, circular in plan,
cylindrical in volume, and with a domelike
roof. He used stones, wood, and mud.
His travails finished, he came back to tell
the group the building he had erected
was in the shape of the Universe and inside
dwelt the Universe's Gods.*

*Then, using a rod he had taken from the
temple, he made a circle around the village
and with the help of others he encircled
it with a high wall built of earth and stones.*

In the center of the village, next to the temple, he erected a large hut which he then covered completely, except for the entrance, with a mound of earth. On top of this mound he vertically placed six large stone slabs. That, he called his home. The others called it the palace.

When he died, his body was laid down inside the hut he had called his house, together with all his belongings, and his son covered the entrance with the large stone slabs he removed from the mound's top. Some people say this was how architecture started. EA, 1976

Emilio Ambasz

Emilio Ambasz, born in Argentina in 1943, studied at Princeton University. He completed the undergraduate program in one year and the following year gained a master's degree in architecture from the same institution. From 1969 to 1976 he was curator of design at the Museum of Modern Art in New York, where he directed and installed numerous influential exhibitions on architecture and industrial design.

Ambasz was president of the Architectural League for two terms, from 1981 to 1985. Between 1966 and 1969 he taught at Princeton University's School of Architecture and was a visiting professor at the Hochschule für Gestaltung in Ulm, Germany. Since then, he has lectured at many major American universities.

Ambasz has been the subject of numerous international publications as well as museum and art gallery exhibitions. Major international publications such as *Domus, ON, Space and Design, Architectural Record,* and *Architecture plus Urbanism* have dedicated special issues to Ambasz's architectural work.

Ambasz also holds a number of industrial and mechanical design patents. Since 1980 he has been chief design consultant for the Cummins Engine Company, which is internationally celebrated for its enlightened support of architecture and design. Ambasz has received numerous industrial design awards.

The Vertebra Chair is included in the design collections of the Museum of Modern Art, New York, and the Metropolitan Museum of Art, New York. The Museum of Modern Art has also included in its design collection Ambasz's 1967 3-D poster Geigy Graphics and his Flashlight.

Selected exhibitions at The Museum of Modern Art
1969 *Paris: May 1968: Posters of the Student Revolt*
1972 *Italy: The New Domestic Landscape: Achievements and Problems of Italian Design*
1973 *A Classic Car: Cisitalia GT, 1946*
Designing Programs / Programming Designs: An Exhibition of Karl Gerstner
1974 *The Architecture of Luis Barragán*
The American National Pavilion at the Venice Biennale
1976 *The Taxi Project: Realistic Solutions for Today*

Selected exhibitions on Emilio Ambasz's works
1983 *Emilio Ambasz: 10 Years of Architecture, Graphic and Industrial Design,* Milan, Madrid, Zurich
1985 *Emilio Ambasz,* The Axis Design and Architecture Gallery, Tokyo
1986 *Emilio Ambasz,* Institute of Contemporary Art of Geneva at Halle Sud, Switzerland
1987 *Emilio Ambasz,* Arc-en-Ciel Gallery at the Center of Contemporary Art, Bordeaux, France
1989 *Emilio Ambasz: Architecture,* The Museum of Modern Art, New York
Emilio Ambasz: Architecture, Exhibition, Industrial and Graphic Design, San Diego Museum of Contemporary Art, Musée des arts décoratifs de Montreal, Akron Art Museum in Ohio, Art Institute of Chicago in Illinois, Laumeier Sculpture Park in St. Louis
1993 Retrospective, Tokyo Station Contemporary Center, Japan
1994 Retrospective, Centro Cultural Arte Contemporaneo in Mexico City, South America, Europe
2009 *In Situ: Architecture and Landscape,* Museum of Modern Art, New York
2010 *Green over Gray,* Grimaldi Forum, Monaco
2012 Comprehensive major retrospective, Museo Nacional Centro de Arte Contemporanea Reina Sofía, Madrid, Spain

Selected publications
1972 Ambasz, Emilio, ed.: *Italy: The New Domestic Landscape: Achievements and Problems of Italian Design.* New York: The Museum of Modern Art.
1976 Ambasz, Emilio, ed.: *The Taxi Project: Realistic Solutions for Today.* New York: The Museum of Modern Art.
1989 Ambasz, Emilio. *The Poetics of the Pragmatic: Architecture, Exhibit, Industrial and Graphic Design.* New York: Rizzoli International Publications.
1993 Ambasz, Emilio. *Inventions: The Reality of the Ideal.* New York: Rizzoli International Publications.
2001 Ambasz, Emilio. *Natural Architecture, Artificial Design.* Milan: Electa.
2004 Irace, Fulvio. *Emilio Ambasz: A Technological Arcadia.* Milan: Skira.
2005 Ambasz, Emilio. *Casa de Retiro Espiritual.* Milan: Skira.

Selected Architectural Awards and Prizes

- 1976 Progressive Architecture Award for the Grand Rapids Art Museum, Michigan
- 1980 Progressive Architecture Award for La Casa de Retiro Spiritual, north of Seville, Spain
- 1983 Annual Interiors Award for the interior of Banque Bruxelles Lambert, Lausanne, Switzerland
- 1985 Progressive Architecture Award for Conservatory at the San Antonio Botanical Center, Texas
- 1986 First Prize and Gold Medal in the closed competition for the *Universal Exhibition of 1992,* Seville, Spain
  Architectural Projects Award for the *Universal Exhibition of 1992,* Seville, Spain
  First Prize in the 1986 closed competition for the Urban Plan for the Eschenheim Tower, Frankfurt, Germany
- 1987 Grand Prize of the International Interior Design Award for the headquarters of Financial Guaranty Insurance Company, New York
  Nominated for Progressive Architecture Award with *Mercedes Benz Showroom*
- 1988 National Glass Association Award for Excellence in Commercial Design for the Conservatory at the San Antonio Botanical Center, Texas
- 1990 Quaternario Award for Conservatory at the San Antonio Botanical Center, Texas
- 2000 Special award by the Japanese Department of Public Works for the Mycal Cultural Center at Shin-Sanda, Japan
  Saflex Design Award for the Mycal Cultural Center at Shin-Sanda, Japan
  Architectural Award by American Institute of Architects/*Business Week* for Fukuoka Prefectural and International Hall, Japan
- 2001 First Prize from the Japanese Institute of Architects for Fukuoka Prefectural and International Hall, Japan
- 2002 American Architectural Award for Monument Towers, Phoenix, Arizona,
- 2007 Medalla Manuel Tolsá by La Universidad Nacional Autónoma de México, Mexico
  Honorary Fellowship at The American Institute of Architects
- 2013 Commendatore d'Italia, Stella d'Oro, by the Italian government
- 2015 Honorary International Fellow at the Royal Institute of British Architects

Selected Design Awards and Prizes

- 1977 *Gold Prize* by the IBD (USA) for the Vertebra Seating System
- 1979 SMAU Prize (Italy) for the Vertebra Seating System
- 1980 Design Excellence Award by the Industrial Designers Society of America (IDSA) for Logotec spotlight range
- 1981 Compasso d'Oro (Italy) for the Vertebra Seating System
- 1983 Design Excellence Award by the IDSA for Oseris spotlight range
- 1984 Jury Special Award by the Tenth Biennial of Industrial Design, Ljubljana
- 1985 Annual Design Review from *Industrial Design* for Cummins N14 Diesel Engine
- 1986 Design Excellence Award by the IDSA for Escargot air filter design for Fleetguard Incorporated
  IDEA Award from the IDSA for the headquarters of Financial Guaranty Insurance Company of New York, New York
  Nominated for American Institute of Architects' Architectural Projects Award with *Mercedes Benz Showroom*
- 1987 Industrial Excellence Award by the IDSA for Saturno modular lighting system
  Nominated for Compasso d'Oro (Italy) and by the IDSA for Flashlight
- 1988 Industrial Excellence Award by the IDSA
- 1989 ID Designer's Choice Award for AquaColor water color set
- 1991 Compasso d'Oro (Italy) for Qualis seating design
- 1992 Awards by the IDSA for Handkerchief Television, Sunstar Toothbrushes, and Soft Notebook
- 1997 Vitruvius Award by the Museo Nacional de Bellas Artes
- 1999 Award for Design Excellence by the IDSA/*Business Week* for Cummins Signature 600 Engine
- 2000 Gold Award for Design Excellence by the IDSA/*Business Week* for Saturno, a street lighting system
- 2001 Compasso d'Oro (Italy) for Saturno
- 2003 Gold in iF Design Award by the International Forum for Design for Stacker chair design
- 2014 Medal of Science from the Institute for Advanced Studies, University of Bologna, Italy

Authors' Biographies

Barry Bergdoll
is Meyer Schapiro Professor of Art History and Archaeology at Columbia University and curator in the Department of Architecture and Design at MoMA. After graduating from Columbia in 1977, he did postgraduate studies at Cambridge University on a Kellett Fellowship until 1979. He returned to Colombia for his PhD, which he completed in 1986. His main field of interest and research lies in nineteenth- and twentieth-century architectural history, theory, and criticism.

Bergdoll curated several important architectural exhibitions, among them *Mies in Berlin, Le Panthéon: Symbole des Révolutions,* and *Les Vaudoyers: Une dynastie d'architectes.* Between 2007 and 2013 he served as Philip Johnson Chief Curator of Architecture and Design at MoMA. Publications include *Bauhaus 1919–1933: Workshops for Modernity* (2009–2010), *Mies in Berlin* (2001), and *European Architecture 1750–1890* (2000).

Peter Buchanan
is an architect born in Malawi and based in London. After completing his studies at the University of Cape Town, he worked as an architect and urban planner in Africa, Europe, and the Near East. Since 1992 he has been working as an exhibition curator, and as a consultant for urban and green design projects.

During the 1980s he was deputy director of *The Architectural Review,* which published many of his writings. Additionally, he curated exhibitions and volumes on topics such as *Renzo Piano Building Workshop: Complete Works* (1993) and *Ten Shades of Green: Architecture and the Natural World* (2005) for the Architectural League of New York. In 2011 he launched an influential and controversial year-long series of essays and events called *The Big Rethink,* in which he started a comprehensive critique of current architecture in the face of global economic and environmental crises.

Kenneth Frampton
is a British architect, critic, and historian who trained at the Architectural Association School of Architecture in London. Frampton is Ware Professor of Architecture at Columbia University's GSAPP. In 1972 he joined the Institute for Architecture and Urban Studies in New York, where he cofounded and edited the magazine *Oppositions.* Over the years he has taught in various capacities at a number of schools, among them the Royal College of Art, London, Princeton University, and the ETH Zurich.

Known for his critical and theoretical writings on twentieth-century architecture, his books include *Modern Architecture: A Critical History* (1980), *Studies in Tectonic Culture* (1995), and *A Genealogy of Modern Architecture: Comparative Critical Analysis of Built Form* (2015).

Peter Hall
is a design writer, as well as being a senior lecturer and head of the Design department at Griffith University Queensland College of Art. Until 2012, he served as senior lecturer at the University of Texas, Austin. He has given numerous seminars on design and mapping in institutes such as the University of Minnesota Design Institute and the Yale School of Art. Since 2006 he has served as vice president of DesignInquiry, a nonprofit educational organization dedicated to research in design subjects.

Since 2000, Hall has been a frequent contributor to the magazine *Metropolis.* Publications include *Tibor Kalman: Perverse Optimist* (2000), *Pause: 59 Minutes of Motion Graphics* (2000), *Else/Where: Mapping New Cartographies of Networks and Territories* (2006), and *Sagmeister: Made you Look* (2009).

Fulvio Irace
is professor of the history of architecture at the Politecnico di Milano, visiting professor at the Accademia di Architettura di Mendrisio, and a member of the academic consulting of the PhD course "History of Architecture and Town Planning" at Politecnico di Torino. He is involved with the scientific committee of the Vico Magistretti Foundation and is one of the founders of the national association of Architectural Archives of Italy. From 2008 to 2009 he was a jury member for the Mies van der Rohe European Prize, and from 2005 to 2009 a member of the scientific committee of the Milan Triennial and curator of its Architecture and Territory sector.

As architectural editor for the publications *Domus* and *Abitare,* he has worked with the most important national and international magazines in the sector. In 2005 he was awarded the InArch–Bruno Zevi Prize for architectural criticism.

Dean MacCannell
is professor emeritus of landscape architecture at University of California, Davis. Educated as an anthropologist at the University of California, Berkeley, he completed his studies with a PhD in sociology at Cornell University. He is a founding member of the International Tourism Research Institute and is associated with the École freudienne de Québec.

MacCannell is the author of numerous articles and works on landscape architecture, community, and culture. His 1976 book entitled *The Tourist: A New Theory of the Leisure Class* is concerned with the development of the sociology of tourism. Other important publications include *Empty Meeting Grounds: The Tourist Papers* (1992) and *The Ethics of Sightseeing* (2011).

Hans Ulrich Obrist
was born in Switzerland, and is a contemporary art curator, critic, and art historian. He is currently artistic director at the Serpentine Gallery, London. Obrist is the author of *The Interview Project,* an extensive ongoing project of interviews. In 1993 he founded the Museum Robert Walser and began to run the *Migrateurs* program at the Musée d'Art Moderne de la Ville de Paris, where he served as a curator for contemporary art. In 1996 he cocurated the first edition of the European biennial of contemporary art Manifesta. In 2009 he was made an Honorary Fellow of the Royal Institute of British Architects (RIBA).

Obrist has lectured internationally at numerous academic and art institutions. While maintaining official curatorial positions, he is also a contributing editor for the magazines *032c, Abitare, Artforum,* and *Paradis.*

Lauren Sedofsky
completed her studies at Sarah Lawrence College, and then, in 1974, obtained a Fulbright scholarship that allowed her to settle in Paris, where she still resides. She has taught at the Ecole Nationale Supérieure de Statistique, and has written film, art, photography, and architecture critiques for the magazine *Artforum.* Among her most outstanding writings is a digest on the pictorial work of Stephen Posen. In addition, she has written the script for the film *Pola X,* directed by Leos Carax.

Michael Sorkin
is distinguished professor of architecture and director of the graduate program in urban design at City College of New York. Previously he was professor of urbanism and director of the Institute of Urbanism at the Academy of Fine Arts in Vienna, and held various teaching commitments as a visiting professor at architectural schools such as Columbia, Yale, Harvard, Cornell, Illinois, and Pennsylvania.

Sorkin is architecture critic for *The Nation* and for ten years held the same position at *The Village Voice.* He is also a contributing editor at magazines such as *Architectural Record* and *Metropolis,* and has published twenty books, including *Variations on a Theme Park* (1997), *Twenty Minutes in Manhattan* (2009), and *New Orleans Under Reconstruction* (2014).

James Wines
is Stuckeman Professor at Penn State University, and founder and president of Sculpture in the Environment (SITE). He was chair of the Environmental Design department at Parsons School of Design from 1984 to 1990, and in the 1990s was dean of architecture at the University of Pennsylvania.

Internationally known for his commitment to the integration and fusion of buildings and landscapes in their surrounding contexts, in recent years he has played an increasingly active role in the international green movement through projects, classes, conferences, and writings, among which are *Architecture as Art* (1980), *De-Architecture* (1987), *Green Architecture* (2000), and *SITE: Identity and Density* (2005).

Index

A Technological Arcadia 15, 151, 178, 302
Aalto, Alvar 13, 184
Aeschylus 13
Aesop 137
Agamennone Light (1985) 228
Alassio, Michele 12,
American Folk Art Museum, New York (1979) 43, 274
Ancient greek theater 13
Ando, Tadao 109, 111, 120, 237, 282
Andre, Carl 146
Anthology for a Spatial Buenos Aires (Ambasz, Emilio; 1966) 151, 238, 246, 276–277, 281
Archetype, prototype, type, stereotype 268
Architecture as Conceptual art 141
architettura radicale 120–121, 270
Archizoom 9
Aristophanes 13
Aymonino, Carlo 138
Bachelard, Gaston 76
Bacon, Francis 207, 288
Banca degli Occhi, Mestre, Italy (2009) 20, 198
Banque Bruxelles Lambert, Lausanne, Switzerland (1981) 49, 111, 303
Barbie Knoll, Pasadena, California (1995) 182–183
Barcelona Pavilion (1929) 8
Baron Edmond de Rothschild Memorial Museum, Ramat Hanadiv, Israel (1993) 46, 174–175
Barragán, Luis 11, 71, 140, 143, 151, 243, 251, 270, 302
Barthes, Roland 104, 117–121
Bayley, Stephen 208–209, 217–218
Beaux-Arts 17, 65
Bellini, Mario 155, 208, 210–211, 216–217, 249–250, 259
Benjamin, Walter 75, 86, 272, 276–277
Bill, Max 243, 272, 281
Black Book 240–241, 269
Borges, Jorge Luis 13, 140, 144, 151, 175–176, 207–208, 211–212, 214–215, 217, 238, 276–277
Boullée, Étienne-Louis 76, 81, 274
Bramante, Donato 13, 80
Bresson, Robert 284–285
Brief Office Seating System (1995) 213
Buchanan, Peter 38–39, 104–105, 107, 111–112, 120, 304
Buckminster Fuller, Richard 208, 241
Buontalenti, Bernardo 66, 77, 250
Carnegie-Mellon Center, Pittsburgh, Pennsylvania (Eisenman, Peter; 1987–1988) 69
Casa Canales, Monterrey, Mexico (1991) 44–45, 79
Casa de Retiro Espiritual *see* La Casa de Retiro Espiritual

Catharijnebaan, Utrecht, Netherlands 178
Center for Applied Computer Research, Mexico City (1975) 19, 79, 139, 146, 273,
Centre Pompidou *see* Hortus conclusus
Cervantes 13
Chan, Eric 214, 217–218, 271
Chávez, César 16
Cirlot, Juan-Eduardo 212, 217
Civitas (Estrada, Martinez, date unknown) 277
Coda: A Pre-Design Condition (Ambasz, Emilio; 1968) 107, 120–121, 277, 278
Coetzee, J. M. 13
Commercial and Residential Development, The Hague, Netherlands (2002) 47
Complejo de Oficinas, La Venta, Mexico (1993) 47, 84–85
Complexity and Contradiction in Architecture (Venturi, Robert; 1965) 10
Containers (Sottsass, Ettore; 1972) 258
Cooperative of Mexican-American Grape Growers, California (1976) 16, 78, 146, 154, 156–157, 159, 273
Cummins Diesel Engine (1982) 221, 271
Curator of Design at the Museum of Modern Art, New York 71, 262, 267, 301
Derrida, Jacques 69, 81, 84, 86
Designers & Books 13, 21
Diffrient, Niels 211, 217
Dinkeloo, John 16
Domestic Landscape *see* Italy: The New Domestic Landscape
Dorsal Office Seating System (1978) 212, 224, 271
Drexler, Arthur 10, 17, 138, 248
Droog Design 208
Durand, Gilbert 179
Durkheim, Emile 112, 120
Eisenman, Peter 10, 66, 69, 84, 140–142, 151, 283
Emilio's Folly: Man is an Island (Ambasz, Emilio; 1983) 67, 79, 84, 154, 160, 161, 168, 175
ENI Headquarters, Rome, Italy (1963) 47, 164
Eschenheimer Tower, Frankfurt, Germany (1985) 41, 78, 177, 178, 303
Estrada, Martínez 137, 151, 277
Euripides 13
Europe/America, Exhibition at the Venice Biennale (1976) 146
Eye Bank, *see* Banca degli Occhi
Fiat Belfiore, Florence, Italy (Nouvel, Jean; 2002) 164
Five Architects (Drexler, Arthur; 1972) 10, 138, 151
Flexibol Pens (1985) 230, 271
Ford Foundation 16, 241
Foster, Norman 72, 150

Foucault, Michel 145, 152, 172, 177, 178, 239
Frampton, Kenneth 84, 266, 267, 277, 304
Frankfurt Zoo, Frankfurt, Germany (1986) 41–42
Freud, Sigmund 11, 113, 282
Friedman, Yona 147
Fukuoka Prefectural International Hall, Fukuoka, Japan (1990/1994) 20, 42, 80, 81, 87–88, 117, 167, 247–248, 285, 289, 303
Gabetti e Isola 164
Gabriel, Peter 285
Gálvez, Delfina 9
Garner, Philippe 209, 217
Geddes, Robert 267
Geigy Graphics Posters (1966) 233, 301
Glasarchitektur (Scheerbart, Paul; 1914) 166
Glass House at New Canaan (Johnson, Philip; 1949) 10
Glory Art Museum, Hsin-Chu, Taiwan (1998) 47
Grand Rapids Art Museum, Grand Rapids, Michigan 20, 65, 79, 146, 165, 303
Graves, Michael 141–142
Grays 10, 66
Green over Gray 40, 148, 167, 239, 272, 302
Grimshaw, Nicholas 150
Gruppo Strum 249, 270
Guattari, Felix 76, 81, 86
Gwathmey-Siegel 141
Handkerchief TV (1990) 215, 232, 272, 303
Heidegger, Martin 67–68, 84, 144
Heizer, Michael 70, 145, 147, 170–171, 286
Hejduk, John 10–11, 141
Herzog & de Meuron 164
Hicks, Sheila 268
History of the southern Spanish house 9
Hitchcock, Henry Russell 14
Hochschule für Gestaltung, Ulm 247, 267, 301
Hokkaido 82, 166, 167
Horeau, Hector 78
Hortus Conclusus, Centre Pompidou, France (1989) 162
Hospital of the Guardian Angel, *see* Ospedale dell' Angelo
House for Leo Castelli, East Hampton, New York (1980) 43, 72, 78, 209
House of Spiritual Retreat *see* La Casa de Retiro Espiritual
Houston Center Plaza, Texas (1982) 45, 79, 80, 178
Hybrid Landscape 151
I Ask Myself (Ambasz, Emilio; 1988) 84, 120–121, 151, 155, 211, 276–277
IBM Portable Desktop (2000) 215
Institute for Architecture and Urban Studies, New York 9, 10, 12, 140, 240, 304
Institute of Contemporary Art at Tokyo Station 216, 302
Interpretation of Dreams (Freud, Sigmund; 1899) 11
Irby, James E. 214–215, 217
Italian anti-Bauhaus 271
Italian neo-avant-garde 9
Italy: The New Domestic Landscape (1972) 9, 138, 151, 210, 217, 248, 251, 253–254, 269–270, 277, 302
Italy: The New Domestic Landscape, Exhibition Catalog (1972) 139
Jacobs House (Wright, Frank Lloyd; 1944) 15
Jameson, Frederic 110
Jencks, Charles 71, 145, 152
Johnson, Philip 10–11, 72, 304
Judd, Donald 146, 286
Kahn, Louis 83, 272, 277
Kar-A-Sutra (Bellini, Mario; 1972) 259
Koningin Julianaplein, The Hague, Netherlands (OMA; 2002) 164
Krauss, Rosalind 70–71, 84, 145, 170
La Casa Curutchet, La Plata, Argentina (Le Corbusier; 1949) 73
La Casa de Retiro Espiritual, North of Seville, Spain (1975) 7, 8, 22, 43–44, 68, 72, 78, 111–112, 141–145, 179, 247, 289, 302–303
Lacan, Jacques 103, 107, 116, 120
Le Corbusier 14, 66, 73, 79, 141, 184, 243, 246, 277, 281
Ledoux, Claude Nicholas 12, 67, 147
Les structures anthropologiques de l'imaginaire: introduction à l'archétypologie générale (Durand, Gilbert; 1960) 179
Lévi-Strauss, Claude 103–104, 106–109, 115–121
LeWitt, Sol 71, 84, 146, 163, 170, 286
Liberty Park 80
Logotec Light (1984) 231, 303
Loos, Adolf 76, 143, 146, 151
Lucile Halsell Conservatory, San Antonio, Texas (1982) 52, 75, 81, 174–175, 185
Lucretius 14, 239, 284, 289
Lumb-R Chair (1981) 213, 271
Magic realism 11
Maki, Fumihiko 117, 277
Maldonado, Tomas 243, 245, 247, 267–277
MAMBA *see* Museum of Modern Art, Buenos Aires
Manhattan, Capital of the XXth Century (Ambasz, Emilio; 1969) 148, 152, 277–278
Manoir d'Angoussart, Charleroi, Belgium (1980) 72–73, 78, 180
Manrique, César 77
Master Plan for Barletta (1997) 41
Matta-Clark, Gordon 74, 165
Max Operative Office Seating System (1996) 213
McLuhan, Marshall 214

Meier, Richard 141–142
Memory palace 8, 12–13
Mendini, Alessandro 108, 120, 155, 183
Mercedes-Benz Showroom, New Jersey (1985) 77–78, 303
Meta architecture 273
Michelangelo 48, 185, 198
Miller, Irwin 244, 271
Miss, Mary 144, 151
Moles, Abraham 267, 277
MoMA Working Paper 269
MoMA, New York see Museum of Modern Art, New York
Montana House see Private Estate, Montana
Monte Albán 16
Monte Carlo 8, 42, 43
Monument Tower Offices, Phoenix, Arizona (1998) 49, 303
Morris, Robert 146
Mumford, Lewis 83
Musée du Quai Branly, Pairs, France (Nouvel, Jean; 2006) 164
Museum of Modern Art, Buenos Aires, Argentina (1997) 164–165, 243
Museum of Modern Art, New York 8–11, 14, 17, 71, 138, 151, 210, 217, 245, 253–254, 262, 267, 269, 277, 283, 288, 301–302
Mycal Cultural and Athletic Centre, Shin-Sanda, Japan (1990) 48
N14 Cummins Diesel Engine (1982) see Cummins Diesel Engine
National Diet Library, Kansai, Japan (1996) 167–169
National Library, Buenos Aires, Argentina 238, 267
Negative utopia 139
Nelson, George 208–210, 217
Neo-Gramscian 270
New messiah of environmental architecture 151
New Orleans Museum of Art, New Orleans, Louisiana (1983) 79
New Town Center, Chiba, Japan (1989) 162–163
Nichii Obihiro Department Store, Hokkaido, Japan (1987) 82, 166
Nishiyachiyo New Town Center, Yachiyo, Japan 78
Notes Toward the Formulation of a Design Discourse (Ambasz, Emilio; 1968) 242, 268, 277
Nouvel, Jean 164, 166, 282
Nuova Concordia Resort, Castellaneta, Italy (1994) 47, 149
Old Port of Monte Carlo, Monaco (1998) 42
OMA 66, 164
On the Nature of Things (De rerum natura) (Lucretius; 1st ct. BC) 14
Oseris Spotlight (1985) 215, 231, 303
Ospedale dell'Angelo, Venice-Mestre, Italy (2008) 152, 186, 198
Paimio Sanatorium (Aalto, Alvar; 1929/1933) 184
Palazzo ENI see ENI Headquarters
Papanek, Victor 208–209, 217–218

Paseo del Lago (2011–2012) 79
Paxton, Joseph 149, 174, 176
Perspecta 12 242, 268, 277
Pesce, Gaetano 142, 151, 164, 210, 250, 259
Phoenix History Museum, Phoenix, Arizona (1989) 46–47, 78
Piazza della Visitazione, Matera, Italy (2009) 48
Piranesi, Giovanni Battista 74, 84
Piretti, Giancarlo 208, 271
Plaza Mayor, Salamanca, Spain (1982) 17, 46, 80, 177
Polyphemus Flashlight (1983) 214–215, 228
Pompidou Centre, Paris, France see Hortus Conclusus
Portman, John 16
Portoghesi, Paolo 71
Portugal, Salas 270
Pratolino 66, 77, 250
Précisions sur un état présent de l'architecture et de l'urbanisme (Le Corbusier; 1930) 277
Princeton University, New Jersey 9, 10, 13, 84, 140, 142, 240, 245, 248–249, 267–268, 276–277, 283, 285, 288, 301, 304
Private Estate, Montana (1991) 74, 180–181
Pro Memoria Garden, Ludenhausen, Germany (1978) 182
Progressive Architecture Award 7, 151, 180, 303
Public park and residences, Monte Carlo, Monaco (1998) 43
Qualis Office Seating System (1989) 213, 272, 303
Residence-au-Lac, Lugano, Switzerland (1983) 235
Rifkin, James 151
Rimini Seaside Development Center, Rimini, Italy (1990) 40–41
Rittel, Horst 268
Roche, Kevin 16
Roman theaters of antique Spain 13
Rosetta stone 8
Rossi, Aldo 138
Sakamoto, Ryuichi 109, 120
San Antonio Botanical Gardens, see Lucile Halsell Conservatory
Sapper, Richard 215–216, 249, 250
Sarfatti, Gino 271
Saturno Street/Highway lighting (1998) 216, 226, 303
Scheerbart, Paul 166
Schlumberger Research Laboratories, Austin, Texas (1982) 75, 77, 79, 116, 146, 147–149, 275
School of Architecture, Princeton University 267–268, 283, 301
Scott, Felicity 270, 277
Sculpture in the expanded field (Krauss, Rosalind; 1979) 84, 145, 170
Sedofsky, Lauren 64–65, 163, 166, 178, 305
Seligman, Werner 141
Serra, Richard 145, 170, 286
Seville Exposition see Universal Exposition

Shakespeare 13
Shikoku Marine Resort Community 79
Shin-Sanda *see* Mycal Cultural and Athletic Centre
Smithson, Robert 70–71, 73, 81, 83–84, 86, 145, 147, 152, 170–172, 178, 286, 288
Soft manifesto 151
Soft Portable Radio/Cassette Player (1990) 215, 232, 272
Soft Series (1990) 215–216, 272
Soft-tech aesthetic 213, 271, 272
Sontag, Susan 252, 284
Sophocles 13
Sorkin, Michael 104, 109, 117, 120, 138, 151–152, 155, 172, 174, 176, 178, 180, 183, 245, 253, 277, 281, 305
Sottsass, Ettore 67, 84, 104, 108, 114–115, 117, 120–121, 155, 167, 168, 174, 176, 183, 209–211, 214, 217, 249–250, 258
Stacker Contract Chair (1998) 213, 303
Sudjic, Deyan 209, 217
Superstudio 9, 258, 270
Supersurface (Superstudio; 1972) 258
Szathmary, Arthur 285
Tabacalera 165
Tafuri, Manfredo 65–66, 72, 75, 83–84, 86, 142, 151
Taut, Bruno 77, 83, 166
Taxi *see* The Taxi Project
Tennis Office Seating System (1997) 213, 224
The Angel's Hospital *see* Ospedale dell'Angelo
The Art of Memory (Yates, Frances; 1966) 9
The Author as Producer (Benjamin, Walter; 1934) 272, 277
The Natural House (Wright, Frank Lloyd; 1954) 14
The Poetics of the Pragmatic (Ambasz, Emilio; 1988) 84, 120, 139, 151–152, 155, 168, 176, 183, 217–218, 277–278
The Taxi Project: Realistic Solutions for Today (1976) 251–253, 262, 270, 302
The Universitas Project: Solutions for a Post-Technological Society (1972) 237, 239–242, 245, 248–249, 252, 269, 277, 283, 286
Thermal Gardens, Sirmione, Italy (1996) 46, 176
Tronti, Mario 270
Turrell, James 12
Union Station, Kansas City, Missouri (1986) 17, 18, 79, 146, 167
Universal Exposition, Seville, Spain (1992) 116, 275
Universitas Project *see* The Universitas Project
Valéry, Paul 238, 284
van der Rohe, Mies 8, 11, 166, 246, 281, 304
Vathorst shopping centre, Amersfort, Netherlands (1999) 172, 173
Vatican Belvedere (Bramante, Donato; 1506) 13, 80
Venice hospital project (Le Corbusier; 1965) 184

Venice-Mestre Hospital *see* Ospedale dell'Angelo
Venturi, Robert 10, 13, 71
Vertair Chair (1991) 213, 224, 271
Vertebra Office Seating System (1974–75) 208, 209, 211, 220, 222, 224, 242, 244, 271, 283, 301, 303
Vertical garden 162
VoX Contract Chair (1996) 213, 225
Wall Houses (Hejduk, John; 1973) 10
Where the Stress Falls (Sontag, Susan; 2001) 284
Whites 10, 11, 66, 141
Williams, Amancio 9, 14, 73, 139, 140, 245, 246, 251, 267, 281, 282
Wines, James 148, 151, 152, 168, 283, 287, 305
Winnisook Lodge, Catskill Mountains, New York (1993) 47, 149, 172, 173
Wölfflin, Heinrich 72, 86
Worldbridge Trade and Investment Center, Baltimore, Maryland (1989) 47, 80, 81, 149, 167
Wright, Frank Lloyd 14, 15, 66, 73, 81, 83, 166, 168, 268, 288
X-Pand Suitcase (1985) 231
Yates, Frances 9
Zevi, Bruno 69, 304

*My quest for the essential in architecture
is not about being simple and light
like a feather; it is about being essential
and concise, like a bird.* EA

EMILIO OF THE GARDEN

EMILIO AMBASZ
EMERGING NATURE
Precursor of Architecture and Design

Contributions by Emilio Ambasz, Barry Bergdoll, Peter Buchanan, Kenneth Frampton, Peter Hall, Fulvio Irace, Dean MacCannell, Hans-Ulrich Obrist, Lauren Sedofsky, Michael Sorkin, James Wines

Editorial team: Maya Rüegg, Manuel Müller
Translations of Fulvio Irace's texts (It–En): Helen Ferguson
Copyediting of the interviews by Sorkin and Wines:
Sarah Quigley
Proofreading: Keonaona Peterson
Design: Integral Lars Müller / Lars Müller and
Esther Butterworth
Production: Martina Mullis
Lithography: prints professional, Berlin, Germany
Printing and binding: Belvédère, Oosterbeek,
the Netherlands
Paper: 150 gsm, GardaGloss

Concept and content of this publication are based on the catalog *Emilio Ambasz – invenciones: arquitectura y diseño*, issued in 2011 by the Museo Nacional Centro de Arte Reina Sofía in Spain on the occasion of an exhibition of the same title.

© 2017 Lars Müller Publishers and Emilio Ambasz

No part of this book may be used or reproduced in any form or manner whatsoever without prior written permission, except in the case of brief quotations embodied in critical articles and reviews.

Lars Müller Publishers
Zürich, Switzerland
www.lars-mueller-publishers.com

ISBN 978-3-03778-526-3

Printed in the Netherlands

Image Credits
For all texts and images unless otherwise specified:
© 2017 Emilio Ambasz

18 bottom Courtesy of Hallmark
19, 58 bottom, 72, 75, 77 Photos by Louis Checkman
24, 31 Photos by Michele Alassio
30, 37 Photos by Fernando Alda
53, 62 Photos by Greg Hursley
85 Photos by Ruyzo Masunaga
89–93, 96 Photos by Hiromi Watanabe
100, 300 Photos by Wade Zimmerman
258 top © Superstudio / Cristiano Toraldo di Francia, 1972
258 bottom Ettore Sottsass Jr, maquette by Michela Bertolini, Stefania Colombo, and Francesca Scalise, Supervisor Prof. G. Ottolini, Politecnico di Milano (1989–1990). Source: Atlas of Interiors (www.atlasofinteriors.polimi-cooperation.org)
© 2017, ProLitteris, Zurich
259 top Photos by Klaus Zaugg © Zaugg's heir — Regione Lombardia / Museo di Fotografia Contemporanea, Milano-Cinisello Balsamo
259 bottom Curatorial Exhibition Files, Exh.#1004. New York, Museum of Modern Art (MoMA), 1972. Photographer: Leonardo LeGrand. Acc. n.: IN1004.41. Digital image © 2017 The Museum of Modern Art, New York / Scala, Florence
277 Photo from Kenneth Frampton's private archive
311 Drawing by James Wines